"Superb.... Weinman has compassionately given Sally Horner pride of place once more in her own life, a life that was first brutally warped by Frank La Salle, and then appropriated by one of the most brilliant writers of the 20th century."
—*Washington Post*

"*The Real Lolita* not only casts Nabokov's most famous novel in new light but also gives Sally the posthumous literary acknowledgment she deserves." —*NPR*

"A riveting blend of true crime, historical investigation, and literary analysis, Sarah Weinman adds another dimension to this already complicated context." —*BuzzFeed*

"Captivating, heartrending... offers both nuanced and compassionate true-crime reportage and revelatory cultural and literary history. It will, quite simply, change the way you think about *Lolita* and 'Lolitas' forever."
—*MEGAN ABBOTT*

"[A] gripping tale of a long-forgotten victim whose ordeal also echoes the more recent cases of Elizabeth Smart and Jaycee Dugard." —*People*

"Utterly engrossing.... Mesmerizing.... We develop boundless compassion for this once little girl, along with a deep empathy and sorrow for the story of her life."
—*Los Angeles Review of Books*

Praise for Sarah Weinman's

THE REAL LOLITA

"An engaging work of literary criticism and a bold act of reclaiming a narrative from the abuser. . . . With excellent reporting and just enough educated guessing to fill in the blanks, Weinman restores a small shred of dignity to the child who unwillingly inspired the literary nymphet whose name became synonymous with sexual precocity."
—*Vulture*, a Best Crime Book of the Year

"Weinman honors the girl whose trauma fell into the shadow of a literary masterpiece, by placing her at the center—drawing her out, through evidence and inference, as a fully fleshed person rather than a cautionary tale, or a piece of inspiration."
—*BuzzFeed*, a Best Nonfiction Book of the Year

"Intriguing. . . . Weinman intelligently enlarges the history of Nabokov's disturbing classic, *Lolita*."
—Esi Edugyan, *New York Times*

"The achievement of [Weinman's] impressive literary sleuthing is to bring to life a girl whose story had been lost."
—*New York Times Book Review*

"Gripping. . . . Glimpses into Nabokov's process will tantalize die-hard fans, and true-crime aficionados will relish Weinman's assiduous reporting."
—*O, The Oprah Magazine*

"Superb. . . . A compelling investigation. . . . Weinman has evocatively reconstructed Sally's nightmare." —*Fresh Air*

"Through careful storytelling, Weinman retrieves Horner from her role as grist in the mill of literary history." —*The New Yorker*

"[Weinman's] real achievement is in evocatively relating the story of a girl who—like her fictional counterpart—was no temptress . . . but the victim of a sexual predator. . . . [She] has brilliantly filled out her subject's ghost." —*Entertainment Weekly*

"Sarah Weinman . . . doubles up on her literary sleuthing. . . . She also unearths plenty of fascinating, often unexpected material. . . . *The Real Lolita* stands out for its captivating mix of tenacious investigative reporting, well-chosen photographs, astute literary analysis, and passionate posthumous recognition of a defenseless child who—until now—never received the literary acknowledgment she deserved." —NPR.org

"Riveting. . . . Scrupulously researched. . . . Nearly 70 years after Sally Horner's death, Weinman's dark and compulsively readable book will make readers aware of the absence of a nearly forgotten girl's voice in discussions of one of the great works of American literature." —*Los Angeles Times*

"Superb. . . . [Weinman] has now become something of a literary detective herself, conducting an investigation into the case she says inspired *Lolita*. . . . Weinman assembles a substantial array of evidence. . . . Weinman has compassionately given Sally Horner pride of place once more in her own life, a life that was first brutally warped by Frank La Salle, and then appropriated by one of the most brilliant writers of the 20th century." —*Washington Post*

"Part true crime story, part literary mystery. . . . Gorgeously written, *The Real Lolita* reads like a novel and will thrill and captivate readers." —PopSugar

"Sarah Weinman unearths the case of Sally Horner, a schoolgirl who was kidnapped in 1948. . . . Weinman argues that the road-trip and school details provided Nabokov with the scaffolding he needed to finish *Lolita*. . . . She's essentially clinched the case." —*The Atlantic*

"Weinman tells Sally's tragic story as it has never been told before, with sensitivity and depth." —*Publishers Weekly*

"A tantalizing, entertaining true-life detective and literary story." —*Kirkus Reviews*

"Spine-straightening. . . . Weinman's sensitive insights into Horner's struggle play in stunning counterpoint to her illuminations of Nabokov's dark obsession and literary daring, and *Lolita*'s explosive impact." —*Booklist*

"*The Real Lolita* is the corrective we never knew we needed, a lively, engrossing work of scholarship that does not diminish Nabokov, but gently insists that we not indulge his trickster ways. Sally Horner matters and, thanks to Sarah Weinman, she and Dolores Haze will be forever linked. Groundbreaking work, a new genre unto itself." —Laura Lippman

"Sarah Weinman delivers a thoroughly riveting and heartbreaking narrative that weaves the very best of true-crime writing with the darker elements of literary inspiration." —Gilbert King

"*The Real Lolita* is a tour de force of literary detective work. Not only does it shed new light on the terrifying true saga that influenced Nabokov's masterpiece, it restores the forgotten victim to our consciousness."

—David Grann

"Compassionate and necessary, Sarah Weinman's *The Real Lolita* is more than a true-crime achievement. It's a literary rescue mission, bringing to life the tragic real-life case that forms the dark heart of Nabokov's classic. You'll never read Lolita the same way again."

—Robert Kolker

THE REAL LOLITA

THE
REAL LOLITA

A Lost Girl, an Unthinkable Crime,
and a Scandalous Masterpiece

SARAH WEINMAN

ecco
An Imprint of HarperCollins Publishers

A hardcover edition of this book was published in 2018 by Ecco, an imprint of HarperCollins Publishers.

FIRST ECCO PAPERBACK EDITION PUBLISHED 2019.

Designed by Suet Yee Chong

The Library of Congress has catalogued a previous edition as follows:

Names: Weinman, Sarah, author.
Title: The real Lolita : the kidnapping of Sally Horner and the novel that scandalized the world / [by Sarah Weinman].
Description: First edition. | New York, NY : HarperCollins Publishers, 2018. | Includes bibliographical references and index.
Identifiers: LCCN 2018006366 (print) | LCCN 2018021107 (ebook) | ISBN 9780062661944 (ebook) | ISBN 9780062861184 | ISBN 9780062661951 | ISBN 9780062661920 | ISBN 9780062661937
Subjects: LCSH: Horner, Sally. | Kidnapping—United States—Case studies. | Child abuse—United States—Case studies. | Captivity—United States—Case studies.
Classification: LCC HV6603.H67 (ebook) | LCC HV6603.H67 W45 2018 (print) | DDC 362.88092 [B] —dc23
LC record available at https://lccn.loc.gov/2018006366

ISBN 978-0-06-266193-7 (pbk.)

19 20 21 22 23   LSC   10 9 8 7 6 5 4 3 2 1

For my mother

You have to be an artist and a madman, a creature of infinite melancholy, with a bubble of hot poison in your loins and a super-voluptuous flame permanently aglow in your subtle spine (oh, how you have to cringe and hide!), in order to discern at once, by ineffable signs—the slightly feline outline of a cheekbone, the slenderness of a downy limb, and other indices which despair and shame and tears of tenderness forbid me to tabulate—the little deadly demon among the wholesome children; *she* stands unrecognized by them and unconscious herself of her fantastic power.

—Vladimir Nabokov, *Lolita*

I want to go home as soon as I can.

—Sally Horner, March 21, 1950

CONTENTS

THE REAL LOLITA

"Had I Done to Her . . . ?"

"Had I done to Dolly, perhaps, what Frank Lasalle,
a fifty-year-old mechanic, had done to eleven-year-old
Sally Horner in 1948?"

—Vladimir Nabokov, *Lolita*

A couple of years before her life changed course forever, Sally Horner posed for a photograph. Nine years old at the time, she stands in front of the back fence of her house, a thin, leafless tree disappearing into the top right-hand corner of the frame. Tendrils of Sally's hair brush her face and the top shoulder of her coat. She looks straight ahead at the photographer, her sister's husband, trust and love for him evident in her expression. The photo has a ghostly quality, enhanced by the sepia color and the blurred focus.

This wasn't the first photograph of Sally Horner that I saw, and I've seen a great many more since. But this is the one I think of the most. Because it's the only photo where Sally has

Florence "Sally" Horner, age nine.

a child's utter lack of guile, without any idea of what horrors lie ahead. Here was evidence of one future she might have had. Sally didn't have a chance to live that one out.

FLORENCE "SALLY" HORNER disappeared from Camden, New Jersey, in mid-June 1948, in the company of a man calling himself Frank La Salle. Twenty-one months later, in March 1950, with the help of a concerned neighbor, Sally telephoned her family from San Jose, California, begging for someone to send the FBI to rescue her. Sensational coverage and La Salle's hasty guilty plea ensued, and the man spent the remainder of his life in prison.

Sally Horner, however, had only two more years to live. And when she died, in mid-August 1952, news of her death reached Vladimir Nabokov at a critical time in the creation of his novel-in-progress—a book he had struggled with, in various forms, for more than a decade, and one that would transform his personal and professional life far beyond his imaginings.

Sally Horner's story buttressed the second half of *Lolita*. Instead of pitching the manuscript into the fire—Nabokov had come close twice, prevented only by the quick actions of his wife, Véra—he set to finish it, borrowing details from the real-life case as needed. Both Sally Horner and Nabokov's fictional creation Dolores Haze were brunette daughters of widowed mothers, fated to be captives of much older predators for nearly two years.

Lolita, when published, was infamous, then famous, always controversial, always a topic of discussion. It has sold more than sixty million copies worldwide in its sixty-plus years of life. Sally Horner, however, was largely forgotten, except by her immediate family members and close friends. They would not even learn of the connection to *Lolita* until just a few years ago. A curious reporter had drawn a line between the real girl and the fictional character in the early 1960s, only to be scoffed at by the Nabokovs. Then, around the novel's fiftieth anniversary, a well-versed Nabokov scholar explored the link between *Lolita* and Sally, showing just how deeply Nabokov embedded the true story into his fiction.

But neither of those men—the journalist or the academic—thought to look more closely at the brief life of Sally Horner. A life that at first resembled a hardscrabble American childhood, then became something extraordinary, then uplifting, and, last of all, tragic. A life that reverberated through the culture, and irrevocably changed the course of twentieth-century literature.

I TELL CRIME STORIES for a living. That means I read a great deal about, and immerse myself in, bad things happening to people, good or otherwise. Crime stories grapple with what causes people to topple over from sanity to madness, from decency to psychopathy, from love to rage. They ignite within me the twinned sense of obsession and compulsion. If these feelings persist, I know the story is mine to tell.

Some stories, I've learned over time, work best in short form. Others break loose from the artificial constraints of a magazine article. Without structure I cannot tell the story, but without a sense of emotional investment and mission, I cannot do justice to the people whose lives I attempt to re-create for readers.

Several years ago I stumbled upon what happened to Sally Horner while looking for a new story to tell. It was my habit then, and remains so now, to plumb obscure corners of the Internet for ideas. I gravitate toward the mid-twentieth century because that period is well documented by newspapers, radio, even early television, yet just outside the bounds of memory. Court records still exist, but require extra rounds of effort to uncover. There are people still alive who remember what happened, but few enough that their recollections are on the cusp of vanishing. Here, in that liminal space where the contemporary meets the past, are stories crying out for greater context and understanding.

Sally Horner caught my attention with particular urgency. Here was a young girl, victimized over a twenty-one-month odyssey from New Jersey to California, by an opportunistic child molester. Here was a girl who figured out a way to survive away from home against her will, who acted in ways that baffled her friends and relatives at the time. We better comprehend those means of survival now because of more recent accounts of girls and women in captivity. Here was a girl who survived her ordeal when so many others, snatched away from their lives, do not. Then for her to die so soon after her rescue, her story subsumed by a novel, one of the most iconic, important works of the twentieth century? Sally Horner got under my skin in a way that few stories ever have.

I dug for the details of Sally's life and its connections to *Lolita* throughout 2014 for a feature published that fall by the Canadian online magazine *Hazlitt*. Even after chasing down court documents, talking to family members, visiting some of the places she had lived—and some of the places where La Salle took her—and writing the piece, I knew I wasn't finished with Sally Horner. Or, more accurately, she was not finished with me.

What drove me then and galls me now is that Sally's abduction defined her entire short life. She never had a chance to grow up, pursue a career, marry, have children, grow old, be happy. She never got to build on the fierce intelligence so evident to her best friend that, nearly seven decades later, she spoke to me of Sally not as a peer, but as a mentor. After Sally died, her family rarely mentioned her or what had happened. They didn't speak of her with awe, or pity, or scorn. She was only an absence.

For decades Sally's claim to immortality was as an incidental reference in *Lolita*, one of the many utterances by the predatory narrator, Humbert Humbert, that allows him to control the narrative and, of course, to control Dolores Haze. Like Lolita, Sally Horner was no "little deadly demon among the wholesome children." Both girls, fictional and real, *were* wholesome children. Contrary to Humbert Humbert's assertions, Sally, like Lolita, was no seductress, "unconscious herself of her fantastic power."

The fantastic power both girls possessed was the capacity to haunt.

I FIRST READ *LOLITA* at sixteen, as a high school junior whose intellectual curiosity far exceeded her emotional maturity. It was something of a self-imposed dare. Only a few months earlier I'd breezed through *One Day in the Life of Ivan Denisovich* by Alexander Solzhenitsyn. Some months later I'd reckon with *Portnoy's Complaint* by Philip Roth. I thought I could handle what transpired between Dolores Haze and Humbert Humbert.

I thought I could appreciate the language and not be affected by the story. I pretended I was ready for *Lolita,* but I was nowhere close.

Those iconic opening lines, "Lolita, light of my life, fire of my loins. My sin, my soul. Lo-lee-ta," sent a frisson down my adolescent spine. I didn't like that feeling, but I wasn't supposed to. I was soon in thrall to Humbert Humbert's voice, the silken veneer barely concealing a loathsome predilection.

I kept reading, hoping there might be some salvation for Dolores, even though I should have known from the foreword, supplied by the fictional narrator John Ray, Jr., PhD, that it does not arrive for a long time. And when she finally escapes from Humbert's clutches to embrace her own life, her freedom is short-lived.

I realized, though I could not properly articulate it, that Vladimir Nabokov had pulled off something remarkable. *Lolita* was my first encounter with an unreliable narrator, one who must be regarded with suspicion. The whole book relies upon the mounting tension between what Humbert Humbert wants the reader to know and what the reader can discern. It is all too easy to be seduced by his sophisticated narration, his panoramic descriptions of America, circa 1947, and his observations of the girl he nicknames Lolita. Those who love language and literature are rewarded richly, but also duped. If you're not being careful, you lose sight of the fact that Humbert raped a twelve-year-old child repeatedly over the course of nearly two years, and got away with it.

It happened to the writer Mikita Brottman, who in *The Maximum Security Book Club* described her own cognitive dissonance discussing *Lolita* with the discussion group she led at a Maryland maximum-security prison. Brottman, reading the novel in advance, had "immediately fallen in love with the narrator," so much so that Humbert Humbert's "style, humor, and sophistication blind[ed] me to his faults." Brottman knew she shouldn't sympathize with a pedophile, but she couldn't help being mesmerized.

The prisoners in her book club were nowhere near so enchanted. An hour into the discussion, one of them looked up at Brottman and cried, "He's just an old pedo!" A second prisoner added: "It's all bullshit, all his long, fancy words. I can see through it. It's all a cover-up. I know what he wants to do with her." A third prisoner drove home the point that *Lolita* "isn't a *love story*. Get rid of all the fancy language, bring it down to the lower [sic] common denominator, and it's a grown man molesting a little girl."

Brottman, grappling with the prisoners' blunt responses, realized her foolishness. She wasn't the first, nor the last, to be seduced by style or manipulated by language. Millions of readers missed how *Lolita* folded in the story of a girl who experienced in real life what Dolores Haze suffered on the page. The appreciation of art can make a sucker out of those who forget the darkness of real life.

Knowing about Sally Horner does not diminish *Lolita*'s brilliance, or Nabokov's audacious inventiveness, but it does augment the horror he also captured in the novel.

WRITING ABOUT VLADIMIR NABOKOV daunted me, and still does. Reading his work and researching in his archives was like coming up against an electrified fence designed to keep me away from the truth. Clues would present themselves and then evaporate. Letters and diary entries would hint at larger meanings without supporting evidence. My central quest with respect to Nabokov was to figure out what he knew about Sally

Horner and when he knew it. Through a lifetime, and afterlife, of denials and omissions about the sources of his fiction, he made my pursuit as difficult as possible.

Nabokov loathed people scavenging for biographical details that would explain his work. "I hate tampering with the precious lives of great writers and I hate Tom-peeping over the fence of those lives," he once declared in a lecture about Russian literature to his students at Cornell University, where he taught from 1948 through 1959. "I hate the vulgarity of 'human interest,' I hate the rustle of skirts and giggles in the corridors of time—and no biographer will ever catch a glimpse of my private life."

He made his public distaste for the literal mapping of fiction to real life known as early as 1944, in his idiosyncratic, highly selective, and sharply critical biography of the Russian writer Nikolai Gogol. "It is strange, the morbid inclination we have to derive satisfaction from the fact (generally false and always irrelevant) that a work of art is traceable to a 'true story,'" Nabokov chided. "Is it because we begin to respect ourselves more when we learn that the writer, just like ourselves, was not clever enough to make up a story himself?"

The Gogol biography was more a window into Nabokov's own thinking than a treatise on the Russian master. With respect to his own work, Nabokov did not want critics, academics, students, and readers to look for literal meanings or real-life influences. Whatever source material he'd relied on was grist for his own literary mill, to be used as only he saw fit. His insistence on the utter command of his craft served Nabokov well as his reputation and fame grew after the American publication of *Lolita* in 1958. Scores of interviewers, whether they wrote him letters, interrogated him on television, or visited him at his house, abided by his rules of engagement. They handed over their questions in advance and accepted his answers, written at leisure, cobbling them together to mimic spontaneous conversation.

Nabokov erected roadblocks barring access to his private life for deeper, more complex reasons than to protect his inalienable right to tell stories. He kept family secrets, quotidian and gargantuan, that he did not wish anyone to air in public. And no wonder, when you consider what he lived through: the Russian Revolution, multiple emigrations, the rise of the Nazis, and the fruits of international bestselling success. After he immigrated to the United States in 1940, Nabokov also abandoned Russian, the language of the first half of his literary career, for English. He equated losing his mother tongue to losing a limb, even though, in terms of style and syntax, his English dazzled beyond the imagination of most native speakers.

Always by his side, aiding Nabokov with his lifelong quest to keep nosy people at bay, was his wife, Véra. She took on all of the tasks Nabokov wouldn't or couldn't do: assistant, chief letter writer, first reader, driver, subsidiary rights agent, and many other less-defined roles. She subsumed herself, willingly, for his art, and anyone who poked too deeply at her undying devotion looking for contrary feelings was rewarded with fierce denials, stonewalling, or outright untruths.

Yet this book exists in part because the Nabokovs' roadblocks eventually crumbled. Other people did gain access to his private life. There were three increasingly tendentious biographies by Andrew Field, whose relationship with his subject began in harmony but curdled into acrimony well before Nabokov died in 1977. A two-part definitive study by Brian Boyd is still the biographical standard, a quarter century after its publication, with which any Nabokov scholar must reckon. And

Stacy Schiff's 1999 portrayal of Véra Nabokov illuminated so much about their partnership and teased out the fragments of Véra's inner life.

We've also learned more about what made Nabokov tick since the Library of Congress lifted its fifty-year restriction upon his papers in 2009, opening the entire collection to the public. The more substantive trove at the New York Public Library's Berg Collection still has some restrictions, but I was able to immerse myself in Nabokov's work, his notes, his manuscripts, and also the ephemera—newspaper clippings, letters, photographs, diaries.

A strange thing happened as I looked for clues in his published work and his archives: Nabokov grew less knowable. Such is the paradox of a writer whose work is so filled with metaphor and allusion, so dissected by literary scholars and ordinary readers. Even Boyd claimed, more than a decade and a half after writing his biography of Nabokov, that he still did not fully understand *Lolita*.

What helped me grapple with the book was to reread it, again and again. Sometimes like a potboiler, in a single gulp, and other times slowing down to cross-check each sentence. No one could get every reference and recursion on the first try; the novel rewards repeated reading. Nabokov himself believed the only novels worth reading are the ones that demand to be read on multiple occasions. Once you grasp it, the contradictions of *Lolita*'s narrative and plot structure reveal a logic true to itself.

During one *Lolita* reread, I was reminded of the narrator of an earlier Nabokov story, "Spring in Fialta": "Personally, I never could understand the good of thinking up books, of penning things that had not really happened in some way or other . . .

were I a writer, I should allow only my heart to have imagination, and for the rest to rely upon memory, that long-drawn sunset shadow of one's personal truth."

Nabokov himself never openly admitted to such an attitude himself. But the clues are all there in his work. Particularly so in *Lolita,* with its careful attention to popular culture, the habits of preadolescent girls, and the banalities of then-modern American life. Searching out these signs of real-life happenings was no easy task. I found myself probing absence as much as presence, relying on inference and informed speculation as much as fact.

Some cases drop all the direct evidence into your lap. Some cases are more circumstantial. The case for what Vladimir Nabokov knew of Sally Horner and when he knew it falls squarely into the latter category. Investigating it, and how he incorporated Sally's story into *Lolita,* led me to uncover deeper ties between reality and fiction, and to the thematic compulsion Nabokov spent more than two decades exploring, in fits and starts, before finding full fruition in *Lolita.*

Lolita's narrative, it turns out, depended more on a real-life crime than Nabokov would ever admit.

OVER THE FOUR OR SO YEARS I spent working on this book project, I spoke with a great many people about *Lolita.* For some it was their favorite novel, or one of their favorites. Others had never read the book but ventured an opinion nonetheless. Some loathed it, or the idea of it. No one was neutral. Considering the subject matter, this was not a surprise. Not a single person, when I quoted the passage about Sally Horner, remembered it.

I can't say Nabokov designed the book to hide Sally from the reader. Given that the story moves so quickly, perhaps an homage to the highways Humbert and Dolores traverse over many thousands of miles in their cross-country odyssey, it's easy to miss a lot as you go. But I would argue that even casual readers of *Lolita*, who number in the tens of millions, plus the many more millions with some awareness of the novel, the two film versions, or its place in the culture these past six decades, should pay attention to the story of Sally Horner because it is the story of so many girls and women, not just in America, but everywhere. So many of these stories seem like everyday injustices—young women denied opportunity to advance, tethered to marriage and motherhood. Others are more horrific, girls and women abused, brutalized, kidnapped, or worse.

Yet Sally Horner's plight is also uniquely American, unfolding in the shadows of the Second World War, after victory had created a solid, prosperous middle class that could not compensate for terrible future decline. Her abduction is woven into the fabric of her hometown of Camden, New Jersey, which at the time believed itself to be at the apex of the American Dream. Wandering its streets today, as I did on several occasions, was a stark reminder of how Camden has changed for the worse. Sally should have been able to travel America of her own volition, a culmination of the Dream. Instead she was taken against her will, and the road trip became a nightmare.

Sally's life ended too soon. But her story helped inspire a novel people are still discussing and debating more than sixty years after its initial publication. Vladimir Nabokov, through his use of language and formal invention, gave fictional authority to a pedophile and charmed and revolted millions of readers in the process. By exploring the life of Sally Horner, I reveal the truth behind the curtain of fiction. What Humbert

Humbert did to Dolores Haze is, in fact, what Frank La Salle did to Sally Horner in 1948.

With this book, Sally Horner takes precedence. Like the butterflies that Vladimir Nabokov so loved, she emerges from the cage of both fiction and fact, ready to fly free.

ONE

The Five-and-Dime

S ally Horner walked into the Woolworth's on Broadway and Federal in Camden, New Jersey, to steal a five-cent notebook. She'd been dared to by the clique of girls she desperately wanted to join. Sally had never stolen anything in her life; usually she went to that particular five-and-dime for school supplies and her favorite candy. The clique told her it would be easy. Nobody would suspect a girl like Sally, a fifth-grade honor pupil and president of the Junior Red Cross Club at Northeast School, to be a thief. Despite her mounting dread at breaking the law, she believed them. She had no idea a simple act of shoplifting on a March afternoon in 1948 would destroy her life.

Once inside Woolworth's, Sally reached for the first notebook she spied on the gleaming white nickel counter. She stuffed it into her bag and walked away, careful to look straight ahead to the exit door. Before she could cross the threshold to freedom, she felt a hand grab her arm.

Sally looked up. A slender, hawk-faced man loomed above her, iron-gray hair underneath a wide-brimmed fedora, eyes shifting between blue and gray. A scar sliced his cheek by the right side of his nose, while his shirt collar shrouded another mark on his throat. The hand gripping Sally's arm bore the traces of an even older, half-moon stamp forged by fire. Any adult would have sized him up as middle-aged, but to ten-year-old Sally, he looked positively ancient.

"I am an FBI agent," the man said to Sally. "And you are under arrest."

Sally did what many young girls would have done in a similar situation: She cried. She cowered. She felt immediately ashamed.

The man's low voice and steely gaze froze her in place. He pointed across the way to City Hall, the tallest building in Camden. That's where girls like her would be dealt with, he said. Sally didn't understand his meaning at first. Then he explained: to punish her for stealing, she would be sent to the reformatory.

Sally didn't know that much about reform school, but what she knew was not good. She kept crying.

Then his stern manner brightened. It was a lucky break for a little girl like her, he said, that he was the one who caught her and not some other FBI agent. If she agreed to report to him from time to time, he would let her go. Spare her the worst. Show some mercy.

Sally stopped crying. He was going to let her go. She wouldn't have to call her mother from jail—her poor, over-worked mother, Ella, still struggling with the consequences of the suicide of her alcoholic husband, Sally's father, five years earlier; still tethered to her seamstress job, which meant that Sally, too often, went home to an empty house after school.

But she couldn't think about that. Not when she was about to escape real punishment. Any desire she felt about joining the girls' club fell away, overcome by relief she wouldn't face a much larger fear.

Sally did not know the reprieve had an expiration date. One that would come due at any time, without warning.

MONTHS PASSED WITHOUT further word from the FBI man. As the spring of 1948 inched its way to summer, Sally finished up fifth grade at Northeast School. She kept up her marks and remained on the honor roll. She also stuck with the Junior Red Cross and continued to volunteer at local hospitals. Her homeroom teacher, Sarah Hanlin, singled Sally out as "a perfectly lovely girl. . . . [A] better than average pupil, intelligent and well behaved." Sally had had a major escape. She must have been grateful for each successive day of freedom.

The Camden of Sally's girlhood was far removed from the Camden of today. Emma DiRenzo, one of Sally's classmates, remembered it as a "marvelous" place to grow up in. "Everything about Camden back then was wonderful," she said. "When you tell people now, they look at you with big eyes." There were pep rallies at City Hall and social events at the YMCA. Girls jumped rope on the sidewalks, near houses adorned with marble steps. Camden residents took pride in their neighborhoods and communities, whether they were among the Italians in South Camden, the Irish in the city's North Side, the Germans in the East Side neighborhood of

Cramer Hill, or the Polish living along Mt. Ephraim Avenue, lining up to buy homemade kielbasa at Jaskolski's or fresh bread at the Morton Bakery. They didn't dream of suburban flight because there was no reason to leave.

Sally lived at 944 Linden Street, between Ninth and Tenth Streets. Cornelius Martin Park lay a few blocks east, and the city's main downtown was within walking distance to the west, and the Ben Franklin Bridge connecting Camden to Philadelphia was minutes away. The neighborhood was quiet but within reach of Camden's bustling core. Now it isn't a neighborhood at all. The town house where Sally grew up was demolished decades ago. What houses remain across the street are decrepit, with boarded-up windows and doors.

Sally's life in Camden was not idyllic. Despite outward appearances, she was lonely. Sally knew how to take care of herself but she wished she didn't have to. She didn't want to come home to an empty house after school because her mother was working late. Sally couldn't help comparing her life with those of her classmates, who had both mother and father. She confided her frustrations to Hanlin, her teacher, who often walked home with her at the end of a school day.

It's not clear if Sally had close friends her age. Perhaps her desire to be accepted by the popular girls stemmed from a lack of companionship. Her father, Russell, had died three weeks before Sally's sixth birthday, and she'd hardly seen him much before then. Her mother, Ella, worked long hours, and was tired and distant when she was at home. Her sister, Susan, was pregnant with her first child. Sally looked forward to becoming an aunt, whatever being an aunt meant, but it made the eleven-year age gap between the sisters all the more unbridgeable. Sally was still a little girl. Susan was not only an adult, but about to be a mother.

SALLY HORNER WAS WALKING home from Northeast School by herself after the last bell on a mid-June day in 1948. The route from North Seventh and Vine to her house took ten minutes by foot. Somewhere along the way, Sally was intercepted by the man from Woolworth's. Sally had dared to think he'd forgotten about her. Seeing him again was a shock.

Keep in mind that Sally had just turned eleven. She believed he was an FBI agent. She felt his power and feared it, even though it was false. She was convinced if she didn't do what he said that she would be sent to the reformatory and be subject to its horrors, as well as worse ones conjured up in her imagination. No matter how he did it, the man convinced Sally that she must go with him to Atlantic City—the government insisted.

But how would she persuade her mother? This would be no easy task, despite Ella's general state of apathy and exhaustion. The man had an answer for that, too. Sally was to tell her mother that he was the father of two school friends who had invited her to a seashore vacation after school ended for the year. He would take care of the rest with a phone call to her mother. Sally wasn't to worry—he would never let on that she was in trouble with the law. He sent the girl on her way.

At home, Sally waited for her mother to return from work, then parroted the FBI man's story. Ella was uneasy, and let it show. Sally sounded sincere in her desire to go to the Jersey Shore for a week's vacation with friends, but who were these people? Ella had never heard Sally mention the names of these two girls before, nor that of their father, Frank Warner. Or if she had, Ella didn't recall.

The telephone rang. The man on the other end of the line told Ella he was Mr. Warner, father to Sally's school friends. His manner was affable, polite. He seemed courteous, even charming. Sally stayed by her mother as the conversation un-folded. "Warner" told Ella that he and his wife had "plenty of room" in their five-room apartment in Atlantic City to put Sally up for the week.

Under the force of his persuasion, Ella let her concerns slide. "It was a chance for Sally to get a little vacation," she said weeks later. "I couldn't afford to give her one." She did wonder why Sally didn't seem to be all that excited about the vacation. It was out of character. Normally her bright little girl loved to go places.

On June 14, 1948, Ella took Sally to the Camden bus de-pot. She kissed her daughter goodbye and watched her climb aboard an express bus to Atlantic City. She spied the outlines of a middle-aged man, the one she took to be "Warner," next to Sally, but he did not come out to greet her. Ella also did not see anyone else with the man, neither wife nor children. Still, she tamped down her suspicions. She wanted so badly for her daughter to enjoy herself. And it seemed, from the first few let-ters Sally sent her from Atlantic City, that the girl was having a good time.

Ella Horner never dreamed that, within weeks, her girl would become a ghost. By sending Sally off on that bus to Atlantic City, she had consigned her daughter to the stuff of nightmares that would rip any mother apart.

TWO

A Trip to the Beach

Robert and Jean Pfeffer were newlyweds who couldn't afford a honeymoon. So the couple, both twenty-two, settled on a day trip with their family, which included Robert's mother, Emily, his seventeen-year-old sister, also named Emily, his nine-year-old younger sister, Barbara, and four other relatives whose names have been lost to time. The sea-cooled Brigantine Beach, a small town east of Atlantic City, was far enough away from their North Philadelphia neighborhood of Nicetown to feel like a treat, but close enough to get back home by nightfall.

Robert, Jean, Emily senior and junior, and Barbara piled into Robert's car on a weekend morning in July 1948 and set out for the beach. (The other four relatives had their own car.) Somewhere along Route 40, a tire blew out. Robert's car went off the road and landed on its side.

The Pfeffers climbed out, shaken and in shock. No one was hurt, thank goodness, but the car was far too damaged to continue. As Robert stood there, wondering how much it would cost to get the car towed and fixed, a station wagon pulled up. A middle-aged man got out of the front seat, and a girl he introduced as his daughter stepped out from the passenger side.

From there the story Robert Pfeffer told both the *Philadelphia Inquirer* and the Camden *Courier-Post* turns strange, riddled with unsolvable inconsistencies. People turn up where they shouldn't. Chronologies bend out of shape. What's clear is that he was so disturbed by what happened that July morning that he alerted law enforcement and, when they didn't listen, the newspapers.

The man told the Pfeffers his name was Frank and that his daughter's name was Sally. (Robert later recalled the man used La Salle as a last name, but it's unclear if that was really the case.) La Salle offered to take the young couple to get help. Robert and Jean agreed. They got into the back of La Salle's station wagon, and La Salle, with Sally beside him, drove them to the nearest roadside phone. Robert called his father and told him about the accident and the Good Samaritan who had come to their aid. He also asked his father to come pick up his wife and daughters.

There was a hamburger joint at the rest stop, and La Salle, Sally, Robert, and Jean stopped for a quick bite to eat. The waitress seemed to be familiar with Frank and Sally, addressing them by name. Robert figured they must be regulars. After the meal, everyone returned to the wreck and La Salle offered to drive the entire family to Brigantine Beach so their day trip wouldn't be spoiled. He also said he would take care of towing and fixing the car. The Pfeffers accepted.

Sally and Barbara, only two years apart in age, hit it off right away. They went swimming together and played on the beach. La Salle told the Pfeffers that he operated a gas station and garage in Atlantic City, that he was divorced, and that Sally lived with him on summer vacations. Sally behaved as if nothing was amiss. She referred to Frank as "Daddy" and treated him with affection. "She told us how good he had been to her," Pfeffer said.

Later that day, Sally suggested that her "dad" could drive her and Barbara back to their place to clean themselves up. They lived on Pacific Avenue in Atlantic City, just ten minutes away by car.

The minutes passed, then became an hour, then an hour and a half. The Pfeffers, waiting at the beach, started to worry. What was taking so long? Robert's father had arrived, and he offered to squeeze everyone into his car and drive into Atlantic City to see what was going on. Why had they let Barbara go off with strangers, even if one of those strangers was a friendly, blue-eyed little girl? Minutes into the drive, they saw La Salle's station wagon coming toward them, with Sally and Barbara sitting together in the backseat.

They headed back to the wrecked car, which La Salle attached to the back of his station wagon. The group, divided between La Salle's vehicle and Robert's father's car, drove to the Atlantic City garage where La Salle claimed he worked and dropped off the damaged vehicle to be fixed. The body shop, Robert noted, was across the street from a New Jersey State Police station.

Before the Pfeffers went back to Philadelphia, Sally invited Barbara to come stay with her for a weekend. La Salle said they'd love to have the girl visit. The family did not take them up on the invite.

Several days later, the Pfeffers would have even more reason to remember their extended encounter with the middle-
aged man and the girl he claimed was his daughter.

EVERY TIME ELLA HORNER began to wonder if she had done
the right thing in sending Sally off to Atlantic City, a letter
or a call—always from a pay phone—arrived to assuage her
guilt and soothe her mind. Sally seemed to be having a swell
time, or so Ella convinced herself. Perhaps she felt some relief, too, at having a reprieve from the expense of feeding and
entertaining her little girl, which stretched her puny paycheck
beyond its limits.

At the end of her first week away, Sally told her mother she
wanted to stay longer so she could see the Ice Follies. Ella re-
luctantly gave permission. After two weeks, Sally's excuses for
staying in Atlantic City grew more vague, but Ella thought her
daughter still sounded well. Then, at the three-week mark, the
phone calls stopped. Ella's letters to her daughter came back
with "return to sender" stamped on the front.

On July 31, 1948, Ella was relieved to receive another letter.
Sally wrote to say she was leaving Atlantic City and going on
to Baltimore with Mr. Warner. Though she promised to return
home to Camden by the end of the week, she added, "I don't
want to write anymore."

At last, something woke up inside Ella's mind. "I don't
think my little girl has stayed with that man all this time of
her own accord." Her sister, Susan, was days away from giving birth. Would Sally really choose to stay away when she was
about to become an aunt? Ella finally understood the horrible
truth. She called the police.

After Detective Joseph Schultz spoke with Ella, he sent

two other Camden detectives, William Marter and Marshall Thompson, to look for Sally in Atlantic City. On August 4, they arrived at the lodging house on 203 Pacific Avenue that Sally gave as the return address on her letters. There they learned from the landlady, Mrs. McCord, that Warner had been living there, and he'd been posing as Sally's father. There were no other daughters, nor was there a wife. Just one little girl, Sally.

The police also learned the man Ella knew as "Mr. Warner" worked at a gas station, and adopted the alias of "Frank Robinson." When the cops went to the gas station, he wasn't there. He'd failed to show up for work and hadn't even bothered to pick up his final paycheck. "Robinson" had disappeared, and so had Sally. Two suitcases remained in their room, as did several unsent postcards from Sally to her mother. "He didn't take any of his or the girl's clothes, either," Thompson told the *Philadelphia Inquirer*. "He didn't even stop long enough to get his hat."

Among the items left behind in the rooming house was a photograph, one that Ella had never seen before. In it, Sally sat on a swing, feet dangling just above the ground, staring directly at the camera. She wore a cream-colored dress, white socks, and black patent

Photograph of Sally discovered at the Atlantic City boardinghouse in August 1948, six weeks after her disappearance.

shoes, and her honey-streaked light brown hair was pulled away from her face. Her eyes conveyed a mixture of fear and a bottomless desire to please. She looked like she wanted to get this moment right, but didn't know what "right" was supposed to be, when everything was so wrong.

It seemed likely that Sally's kidnapper was the photographer. She was only three months past her eleventh birthday.

Marshall Thompson led the search for Sally in Atlantic City. When that search turned up empty, he took the photo of her back to Camden police headquarters to be sent out on the teletypes. He had to find Sally, the sooner, the better, because police now knew who they were dealing with.

For Sally's mother, it was awful enough that the Camden police had failed to bring her daughter home. Far worse was the news they broke to Ella: the man who had called himself "Warner" was well-known to local law enforcement. They knew him as Frank La Salle. And only six months before he'd abducted Sally, he had been released from prison after serving a sentence for the statutory rape of five girls between the ages of twelve and fourteen.

THREE

From Wellesley to Cornell

The year 1948 was a pivotal one for Vladimir Nabokov. He had spent six years in Cambridge, Massachusetts, teaching literature to Wellesley College undergraduates and, in his spare time, indulging his passion for studying butterflies at Harvard's Museum of Comparative Zoology. After eight years in the United States, the tumult and trauma of emigration had receded. English, Nabokov said many times, was the first language he remembered learning, and the lure of America had sustained him as he fled the Russian Revolution for Germany, and then from the Nazis to Paris—a necessary step when married to a woman who was proud and unafraid to be Jewish.

The United States, and particularly the Boston area, proved a generally happy environment for Nabokov, Véra, and their son, Dmitri, who was fourteen years old in 1948. Since they'd found a haven there, Nabokov had worked on a book about Nikolai Gogol, about whom he had decidedly mixed feelings; published a novel, *Bend Sinister;* and begun the version of his autobiography that would appear as *Conclusive Evidence* a couple of years later. (He would later rewrite it and publish it under the title *Speak, Memory.*)

Nabokov had also traveled across America three times, in the summers of 1941, 1943, and 1947. (He would repeat the cross-country trip four more times.) He never drove, entrusting the task to his wife, Véra, or a graduate student. The first time, Dorothy Leuthold, a middle-aged student in his language class, had spirited the Nabokovs from New York City in a brand-new Pontiac (dubbed Pon'ka, the Russian word for "pony") all the way to Palo Alto, California.

The trio stayed in motor courts and budget hotels and other cheap lodgings that wouldn't break the bank. The America Nabokov witnessed on these trips was eventually immortalized as the "lovely, trustful, dreamy, enormous country" that Humbert Humbert comments on in *Lolita:* "Beyond the tilled plain . . . there would be a slow suffusion of inutile loveliness, a low sun in a platinum haze with a warm, peeled-peach tinge pervading the upper edge of a two-dimensional, dove-gray cloud." Though his marriage to Véra was once again stable, an affair had nearly derailed it a decade earlier, when she had gone on to Paris before him. Perhaps news of his romantic attentions to at least one Wellesley student had not reached Véra—or if it had, she did not view the dalliance as anything serious.

Nabokov had been ill for much of the first half of 1948. He suffered a litany of lung troubles during the spring that no doctor could adequately diagnose. They thought it might be tuberculosis because of the alarming quantities of blood Nabokov coughed up. It wasn't. The next guess was cancer. That, too, proved untrue. When doctors put a vulcanized rubber tube down his windpipe under local anesthetic to inspect his ailing lungs, all they found was a single ruptured blood vessel. Nabokov himself figured his body was "ridding itself of the damage caused by thirty years of heavy smoking." Bedridden, he had enough energy to write, but not to teach, so Véra stood in for him as lecturer.

After these summer trips, Nabokov was always glad to return to Cambridge. Wellesley, his academic and personal refuge, had turned down his multiple entreaties for a full-time professorship. Nor could he find full-time work at Harvard, where he'd made a quixotic bid to turn his butterfly-hunting hobby into a proper profession. But the Nabokovs' fortunes were about to change thanks to Morris Bishop, a romance literature professor at Cornell who would remain a close friend to both Vladimir and Véra. Bishop lobbied Cornell to appoint Nabokov a professor of Russian literature, and it worked. On July 1, the Nabokovs moved to Ithaca, New York, finding solace in a "quiet summer in green surroundings." By August, they had rented a large house on 802 East Seneca Street, one far bigger than their "wrinkled-dwarf Cambridge flatlet"—and future inspiration for the house where a man named Humbert Humbert would discover the object of his obsession.

The summer also brought Nabokov a formative book, thanks to the literary critic Edmund Wilson, who sent Nabokov a copy of Havelock Ellis's *Studies in the Psychology of Sex*. He drew attention to one appendix that contained the late-nineteenth-century confession of an unnamed engineer of Ukrainian descent. The man had first had sex at age twelve with another child, found the experience so intoxicating he repeated it, and eventually destroyed his marriage by sleeping with child prostitutes. From there the man went further downhill, to the point where he

flashed young girls in public. The confession, as Nabokov related in a later interview, "ends with a feeling of hopelessness,
of a life ruined by hunger beyond control."

Nabokov appreciated Wilson's gift and wrote him after
reading the case histories. "I enjoyed the Russian's love-life
hugely. It is wonderfully funny. As a boy, he seems to have
been quite extraordinarily lucky in coming across [willing
girls]. . . . The end is rather bathetic." Nabokov also directly
acknowledged the impact of Ellis to his first biographer. "I was
always interested in psychology," he told Andrew Field. "I knew
my Havelock Ellis rather well. . . ."

He was, by this point, five years from finishing the manuscript for *Lolita,* and a decade from its triumphant American
publication. But Nabokov was also nearly twenty years into his
efforts to wrestle a thematic compulsion into its final form: the
character who became Humbert Humbert.

SKIP PAST THE OFT-QUOTED opening paragraph of *Lolita*'s
first chapter. Chances are, even if you've never read the novel,
you probably know it by heart, or some version of it. Move
directly to paragraph two: "She was Lo, plain Lo, in the morning, standing four feet ten in one sock. She was Lola in slacks.
She was Dolly at school. She was Dolores on the dotted line."

In Humbert Humbert's eyes, the girl named Dolores Haze
is a canvas blank enough to project whatever he, and by virtue
of his narration, the reader, sees or desires—"But in my arms
she was always Lolita." She is never allowed to be herself. Not
in Humbert's telling.

When the reader meets her, Dolores Haze is just shy of
twelve years old, born around the first of the year in 1935, making her two years and three months older than Sally Horner.

She is an inch shorter than Sally and, at seventy-eight pounds, a good twenty pounds lighter than her real-life counterpart. There aren't other facts and figures available for Sally, but Humbert measures every physical aspect of Dolores: twenty-seven-inch chest, twenty-three-inch waist, twenty-nine-inch hips, while her thigh, calf, and neck circumferences were seventeen, eleven, and eleven, respectively.

Dolores's mother, the former Charlotte Becker, and her father, Harold Haze, were living in Pisky, a town somewhere in the Midwest best known for producing hogs, corn, and coal, when their daughter was born. Conception, however, took place in Veracruz, Mexico, during the Hazes' honeymoon. Another child followed in 1937, the year of Sally's birth, but that offspring, a blond-haired boy, died at two. Sometime thereafter—Humbert is vague on details—Harold also perished, leaving Charlotte a widowed single mother. She and Dolores move east to Ramsdale, and set up house at 342 Lawn Street, where both mother and daughter will encounter a man who will alter their lives irrevocably and with monumental consequences.

When he first sees her, Humbert Humbert describes Dolores in poetic terms: "frail, honey-hued shoulders . . . silky supple bare back . . . chestnut head of hair" and wearing "a polka-dotted black kerchief tied around her chest" that shields her breasts from Humbert's "aging ape eyes."

Humbert confides to the reader that when he was nine, he met a girl named Annabel Leigh, also nine. They embarked on a friendship with strong romantic overtones and multiple rendezvous by the beach. Then Annabel fell ill and died prematurely, the idyll forever cut short. Her death imprinted a type, and a predilection, upon Humbert for the rest of his days. Girls who fall between the ages of nine and fourteen. Girls whose "true nature," according to Humbert, bore little resemblance to real life. Girls he characterized as "little deadly demons." Girls immortalized, forevermore, by him as well as his creator, as nymphets.

Humbert Humbert was describing a compulsion. Vladimir Nabokov set out to create an archetype. But the real little girls who fit this idea of the mythical nymphet end up getting lost in the need for artistic license. The abuse that Sally Horner, and other girls like her, endured should not be subsumed by dazzling prose, no matter how brilliant.

FOUR

Sally, at First

The seeds of Sally Horner's kidnapping grew out of choices made by her mother. Ella kept secrets about the circumstances of her daughters' births and the death of Sally's father. Sally never knew of them. Susan may have, but if so she never spoke of them to her family. Digging up these secrets transformed me into an accidental forensic genealogical detective. I spent so many months stuck on Sally's origin story and the clash between what was reported and what really happened because I thought it would help me better understand Ella's behavior.

Her decisions, with respect to Sally's disappearance, hold her up to severe scrutiny by the modern world. She let her daughter go off with a stranger she'd only spoken to by telephone. She grew more distant, perhaps more baffling, to her family, let alone to neighbors. She fit the pejorative "difficult" bill so often affixed to women who don't fit within neat little boxes. But in 1948, with little money and fewer resources available to Ella as a single mother, she functioned within her own limited framework. Her best was not good enough for Sally, but it was all she knew based on the life she'd lived up until she saw her daughter off at the Camden bus depot.

SALLY WAS HER NICKNAME. No one is alive to remember why, or who used it first, or how it stuck. Her legal name, listed on the certificate announcing her birth at Trenton Hospital on April 18, 1937, was Florence, no middle name, Horner. Her mother, the former Ella Katherine Goff, took the baby back to the home she shared with Russell Horner in Roebling, New Jersey. The house at 238 Fourth Avenue is long gone, replaced by a more modern town house a stone's throw from the River Line train station to the east, and several blocks south of the Delaware River.

Ella's older daughter, Susan, also lived with the Horners, though Russell was not her father. Eleven years earlier, at the age of nineteen, Ella had had some sort of relationship with an older man of about thirty. When the subject came up, Ella told her family that she and Susan's birth father, whom she never named, were married, but that he passed away. Susan knew her father's real name, William Ralph Swain, because she listed it on her marriage license. She likely knew little else.

Ella had good reason to keep Swain's existence a secret, and never mention him by name. Records indicate he was married to someone else when Susan was born, contrary to the "yes" ticked off on Susan's birth certificate, indicating her legitimate status. Nor could I find any existing marriage record between Swain and Ella, though one may turn up in the future—vital records are irregularly stored from city to city, state by state. To confuse matters further, the 1930 census listed Ella's last name as Albara, which she used for at least a half dozen more years. The census record also listed her as being married, but I could not track down any marriage record between Ella and a man named Albara.

Ella raised Susan on her own, with occasional help from her parents, Job and Susannah Goff. One subject they all fretted about was how long it took Susan to learn to speak. She'd had some sort of head injury as a baby, and did not begin talking in earnest until she was five, by which time she and her mother had moved to Prospertown to be closer to Ella's parents, who were growing older and more infirm.

That's where Ella met Russell Horner, a widower with a son, also named Russell. Horner began to court her, and some of their meetings were recorded by the local papers, as was the custom of the day. On December 9, 1935, the *Asbury Park Press* noted that Ella and Russell were "recent visitors to friends in Lakehurst." The paper also reported on June 8, 1936, that Ella and Susan visited Russell and his son (both names were spelled as "Russel") in New Egypt, and noted a solo visit by Ella to the town on August 8. Ella and Russell were not husband and wife, though. It seems Ella had repeated the pattern begun with Swain. While Russell's first wife, the mother of his son, had died, he had married a second time and never bothered to divorce the woman. By the end of 1937, Russell and Ella were living as husband and wife at the Fourth Avenue house in Roebling.

As for Russell Junior, he married two months before Sally Horner's birth. Sally never knew of her half brother's existence. Neither her mother nor Susan mentioned him.

Ella and Russell's domestic arrangement was as short-lived

as their earlier relationships. By the time Sally was about three, the situation had grown volatile. Russell had a drinking problem, which did not mix well with his job as a crane operator, and he could be abusive to his wife and her daughters. Susan remembered the beatings her stepfather gave her mother, memories she did not allow herself to think about until close to the end of her own life. Sally, much younger when her parents split up, may have been spared the worst of these memories.

Eventually, Ella fled her relationship. She took Susan and Sally to Camden, where they moved into the town house at 944 Linden Street. Russell became itinerant, drifting from town to town around southern New Jersey, looking for and not finding work. He lost his driver's license when caught taking a shortcut along some railroad tracks. By the beginning of 1943, he was living at his parents' farm in Cassville. On March 24, he hanged himself from the rafters of the garage. Horner left a note for his mother in the kitchen, directing her where to find his body. According to the state police, he "had been despondent over ill health for some time."

Police told the *Asbury Park Press* that Russell had been married twice and was "estranged from his current wife," though they did not say whether the wife in question was Ella. But the address listed on Russell's death certificate was the address where he'd lived with Ella and the girls in Roebling. And the name of his daughter, "Florence," is handwritten just below the address line.

Sally was not quite six years old when her father committed suicide. It isn't clear how much she knew of her father's history and manner of death. Later, when it became necessary to clarify her parentage, she said, "My real daddy died when I was six and I remember what he looks like."

After Russell killed himself, Ella, already living as a single mother, was truly on her own. Her mother, Susannah, had passed away in 1939, while her father, Job, died in January 1943, just two months before Russell killed himself. Ella had to go to work as a seamstress.

Susan, by now sixteen, had left school and was working a factory job. That summer, Susan met Alvin Panaro, a sailor on leave, at a friend's party. Though Al was three years older, he was immediately smitten, but the Second World War was on and they were too young to marry. Al hailed from Florence, near the same part of town where Susan and Sally had once lived. His parents owned a greenhouse, and planned for Al to take over responsibility running it once he was finished with the navy, once he was home for good. It was also understood he might not get that chance; even the navy carried a high casualty risk.

When Susan turned eighteen, they decided not to wait any longer. On his next furlough, she and Al wed in Florence on February 17, 1945. When the war ended, Al received his honorable discharge and he and Susan began married life in earnest, running the greenhouse together. They wanted children, but Susan's initial pregnancies ended in early miscarriages. Then their luck turned.

In June 1948, Susan and Al Panaro were two months away from the birth of their daughter, Diana, Ella's first grandchild. But when the baby girl arrived that August, celebration was the furthest thing from the minds of her parents and grandmother. Sally had disappeared and they knew who had taken her. They also now knew what sort of man he was.

FIVE

The Search for Sally

An eight-state police search for Sally Horner began on August 5, 1948. By then she had been gone from Camden for six weeks. The news wires picked up the story of her abduction, as well as Ella's delay in reporting her daughter missing. The picture of Sally on the swing went out across the country, appearing in wire reports published from Salt Lake City, Utah, to Rochester, New York, and in local papers like the Camden *Courier-Post* and the *Philadelphia Inquirer*.

Robert and Jean Pfeffer were among those who read the news about Sally Horner's disappearance. How strange, the couple thought. "If [Sally] had wanted to warn us about anything she had every opportunity, but never did so." Robert picked up the phone, called the Camden police, and told the officer who answered about their encounter in Brigantine Beach. Robert also mentioned his little sister Barbara's visit to

La Salle's apartment, which had stretched to ninety agonizing minutes of waiting. Perhaps reading about La Salle's prior incarceration made him wonder what, exactly, might have happened to Barbara during those ninety minutes. He never heard a word back from the police.

The shock of the news about Sally, combined with mundane family matters, delayed Pfeffer from making the two-and-a-half-hour round trip back to Atlantic City to pick up their car for several weeks. He never learned whether La Salle himself or some other mechanic restored it to good working order.

Sally and La Salle, however, were long gone from Atlantic City. Camden police now knew, with queasy certainty, why Sally's family had ample reason to be fearful of what Frank La Salle might do to their little girl.

AT FIRST MARSHALL THOMPSON worked the Sally Horner case with other Camden police officers. But when the summer of 1948 gave way to fall, he took on the investigation full-time and never stopped. As the months wore on, his colleagues weren't shy about voicing their opinions. The girl had to be dead. She couldn't up and vanish like this, no trace, no word, when they knew who had her, what they both looked like, and that they were posing as father and daughter.

Thompson felt otherwise. Sally must be alive. He figured it was likely she was still near enough to Camden. And even if she wasn't, he would find her. It was his job as detective to care about every case, but the plight of a missing girl really got to him.

He had been promoted to detective only the year before, nearly two decades into his time on the force. Thompson's appointment in March 1928 happened the same year his only daughter, Caroline, was born, and not long after he and his wife, Emma, moved to the Cramer Hill neighborhood in Northeast Camden. The young couple had long-standing Camden roots, Thompson in particular. His father, George, had served as justice of the peace, and his grandfather John Reeve Thompson was a member of Camden's first city council.

Tangles with "local pugilists," raids on illegal speakeasies, breaking up home gambling dens, and other minor crimes littered Thompson's stretch as a Camden cop. Most of the time he worked with Sergeant Nathan Petit; their names often appeared together in the local papers' accounts of various notable arrests.

Off duty, Thompson entertained family and friends by playing classical piano, which his mother, Harriette, taught him as a child. His musical ability was called out with hyperbolic flourish by a *Courier-Post* columnist in 1939: "Marshall Thompson, one of Camden's finest, is a talented pianist. He never took a music lesson."

Thompson's innate tenacity made him the perfect choice to look for Sally Horner and Frank La Salle. Over the course of his investigation Thompson learned much about Sally's abductor, from his choice of haircut to the "quantity of sugar and cream he desired when drinking coffee." He chased every lead and followed up on every tip. One phone call came in to say La Salle was holed up in a house on Trenton Avenue and Washington Street in downtown Camden. A state police teletype arrived placing La Salle at a residence on Third and Sumner Avenue in Florence, the same town where Sally lived as a little girl. Neither tip panned out.

Once Thompson was on the case full-time, he let it dictate his entire waking life. He got in touch, in person and by telephone, with the FBI; state and city police at Columbus,

Newton, Riverton, and Langhorne, Pennsylvania; state parole offices at Trenton and Camden; detective divisions in Philadelphia; and the Trenton post office. Several months into Sally's disappearance, Thompson received reports of La Salle being spotted in Philadelphia, northern New Jersey, South Jersey Shore resorts, and at a restaurant in Haddonfield, observed by a waitress working there. He followed each lead to no avail.

Thompson also cast his net farther and deeper in the surrounding states. He checked in regularly with state and city police in Absecon, Pleasantville, Maple Shade, Newark, Orange, and Paterson, New Jersey; parole offices in Atlantic City and the state prison farm in Leesburg; and the Compensation Bureau in Trenton, in case La Salle drew or cashed a paycheck in the state.

There were periods where he worked twenty-four hours or more without taking a break. Finding Sally Horner was more important than sleep. Thompson tracked down La Salle's first wife in Portland, Maine, but she knew nothing of his whereabouts. The detective also contacted La Salle's second wife, now living in Delaware Township with her daughter, her new husband, and their baby son. The woman gave Thompson an earful about her wayward, criminal ex-husband's habits and history, including the dramatic beginning of their marriage and its equally explosive end.

Thompson used his holidays to travel for the case. On one six-day "vacation," Thompson went to the Trenton State Fair. Each morning, he stood outside the entrance to the grounds, hoping that La Salle might turn up to apply for a job. Or perhaps he would bring Sally with him.

None of the leads amounted to anything. Nor did tips from numerous anonymous phone calls and letters. All had to be followed up on, but none yielded the answer Detective Marshall Thompson craved: the whereabouts of Sally Horner.

It was the nature of a detective's job to get hopes up and have them crushed. So many of his colleagues believed the girl was dead. But not Thompson. He could not give up. He knew in his bones that he would, someday, find Sally alive and bring her back to Camden, to her mother, her family.

And that he would find Frank La Salle and see justice done.

SIX

Seeds of Compulsion

*Vladimir Nabokov holding a butterfly, 1947, at Harvard's
Museum of Comparative Zoology, where he was a fellow.*

A s Marshall Thompson continued to track Frank La
Salle's whereabouts without results, Vladimir Nabokov remained on a quest to plumb the fictional mind
of a man with a similar appetite for young girls. So far, he had
not been successful. He could have, and tried to, abandon it
altogether—there were plenty of other literary projects for

Nabokov to pursue. But the drive to get this story right went beyond formal exercise. Otherwise, why did Nabokov explore this same topic, over and over, for more than twenty years? At almost every stage of his literary career, Nabokov was preoccupied with the idea of the middle-aged man's obsession with a young girl.

As Martin Amis wrote in a 2011 essay for the *Times Literary Supplement,* "Of the nineteen fictions, no fewer than six wholly or partly concern themselves with the sexuality of prepubescent girls. . . . [T]o be clear as one can be: the unignorable infestation of nymphets . . . is not a matter of morality; it is a matter of aesthetics. There are just too many of them."

"Aesthetics" is one way to phrase it. Robert Roper, in his 2015 book *Nabokov in America,* suggested a more likely culprit: compulsion—"a literary equivalent of the persistent impulse of a pedophile." Over and over, scholars and biographers have searched for direct connections between Nabokov and young children, and failed to find them. What impulses he possessed were literary, not literal, in the manner of the "well-adjusted" writer who persists in writing about the worst sort of crimes. We generally don't bear the same suspicions of writers who turn serial killers into folk heroes. No one, for example, thinks Thomas Harris capable of the terrible deeds of Hannibal Lecter, even though he invented them with chilling psychological insight.

Nabokov likely realized how often this theme persisted in his work. That would explain why he was quick to deny connections between *Lolita* and real-life figures, or to later claim the novel's inspiration emerged from, of all things, a brief article in a French newspaper about "an ape in the Jardin des Plantes who, after months of coaxing by a scientist, produced the first drawing ever charcoaled by an animal: this sketch showed the bars of the poor creature's cage."

But there is no getting around the deep-seated compulsion that recurs again and again in Nabokov's work. I read through his earlier Russian-language novels, as well as more contemporary accounts by literary critics, to figure out why this awful subject held such allure for him.

NABOKOV'S INITIAL EXPLORATION of an older man's unnatural desire for a preteen girl was published in 1926, within the first year of his career as a prose writer. Before then, he devoted himself exclusively to poetry. Did prose free Nabokov up to wrestle with the darkness and tumult that already surrounded him? His father, the jurist and journalist Vladimir D. Nabokov, had been assassinated four years earlier, and he was a year into his marriage to Véra Slonim, a fellow émigré he met while both lived in Berlin among the community of other Russians who'd fled the Revolution. Neither particularly cared for the city, but they stayed in Berlin for fifteen years, Nabokov supplementing his writing income and growing literary reputation by teaching tennis, boxing, and foreign languages to students.

Nabokov published his first novel, *Mashen'ka* (*Mary*), in 1926, under the pseudonym of V. Sirin, which he would use for all of his poetry and prose published before he moved to America. That same year Nabokov, as Sirin, published "A Nursery Tale." The short story includes a section on a fourteen-year-old girl clad in a grown-up cocktail dress designed to show off her cleavage, though it isn't clear that the narrator, Erwin, immediately notices that aspect:

"There was something odd about that face, odd was the flitting glance of her much too shiny eyes, and if she were not just a little girl—the old man's granddaughter, no doubt—one might suspect her lips were touched up with rouge. She walked

swinging her hips very, very slightly, her legs moved closer together, she was asking her companion something in a ringing
voice—and although Erwin gave no command mentally, he
knew that his swift secret wish had been fulfilled."

Erwin's "swift secret wish" is his inappropriate desire for
the girl.

Two years later, in 1928, Nabokov tackled the subject in poetry. "Lilith" also strongly features the so-called demonic effect
of a little girl, of her "russet armpit" and a "green eye over her
shoulder" upon an older man: "She had a water lily in her curls
and was as graceful as a woman." The poem continues:

> And how enticing, and how merry,
> her upturned face! And with a wild
> lunge of my loins I penetrated
> into an unforgotten child.
> Snake within snake, vessel in vessel
> smooth-fitting part, I moved in her,
> through the ascending itch forefeeling
> unutterable pleasure stir.

But this illicit coupling is the man's ruin. Lilith closed herself off to him and forced him out, and as he shouts, "let me
in!" his fate is sealed: "The door stayed silent, and for all to see /
writhing with agony I spilled my seed / and knew abruptly that
I was in Hell." Two and a half decades before *Lolita*, Nabokov
anticipated Humbert Humbert's remark that he was "perfectly
capable of intercourse with Eve, but it was Lilith he longed for."

Another proto-nymphet appears in *Laughter in the Dark*,
though this one, Margot, is a little older: eighteen in the original version published in Russian, *Camera Obscura* (1932), and
sixteen in the heavily revised and retitled edition Nabokov re-

leased six years later. (Nabokov rewrote the novel a third time in the 1960s.) Margot attracts the attention of the much-older, wealthy art critic Albert Albinus,* whose name foreshadows Humbert Humbert.

We only see Margot's actions and personality filtered through Albinus's eyes. He depicts her as capricious, whimsical, and full of manipulation. Just as in *Lolita,* when Humbert's plans are upended by the arrival of Clare Quilty, an interloper foils the relationship between Albinus and Margot. Axel Rex's affair with Margot in *Laughter in the Dark* serves a more mercenary purpose—gaining access to Albinus's status and fortune—while Quilty is after Dolores for the same illicit reasons as Humbert Humbert.

Except for Margot, who is a proper character, the early precursors to Dolores Haze are merely images that tempt and torment Nabokov's male protagonists. The image grows in substance in tandem with Nabokov's artistic growth. A paragraph in *Dar,* written between 1935 and 1937 but not published until 1952 (the English translation, published as *The Gift,* appeared a decade later), all but summarizes the future plot of *Lolita.* "What a novel I would whip off!" declares a secondary character, contemplating his much, much younger stepdaughter:

> Imagine this kind of thing: an old dog—but still in his prime, fiery, thirsting for happiness—gets to know a widow, and she has a daughter, still quite a little girl—you know what I mean—when nothing is formed yet but she has a way of walking that drives you out of

---

* In the original Russian, Albinus was called Bruno Kretchmar, while Margot's name was Magda.

your mind—A slip of a girl, very fair, pale with blue under the eyes—and of course she doesn't even look at the old goat. What to do? Well, not long [after] he ups and marries the widow. Okay. They settle down, the three of them. Here you can go on indefinitely—the temptation, the torment, the itch, the mad hopes . . .

Nabokov did not exactly "whip off" the novel that became *Lolita*. There was one more abortive attempt written in his mother tongue, *Volshebnik*, which was the last piece of fiction he wrote in Russian. He worked on it at a critical point in his life, while waiting to see if he and his family would be able to flee Europe and immigrate to America. But *Volshebnik* would not see publication until almost a decade after his death.

WHEN GERMANY DECLARED WAR on Poland in September 1939, plunging the rest of the world into global battle, Vladimir Nabokov was under considerable stress. He had reunited with his wife, Véra, and their son, Dmitri, in Paris, after an extended separation stranded them in Germany. He had broken off his affair with fellow émigré Irina Guadanini to join his family, but Paris was no safe haven anymore, as the Vichy regime became increasingly close with the Nazis. Véra was Jewish, and so was Dmitri, and if they could not get out of France, they might be bound for concentration camps.

The personal stakes were never higher, and Nabokov's health suffered. That fall, or perhaps in the early winter of 1940, he was "laid up with a severe attack of intercostal neuralgia," a mysterious ailment of damage to nerves running between the ribs that would plague him off and on for the rest of his life. He could not do much more than read and write, and he retreated into the refuge of his imagination. What emerged was *Volshebnik,* the fifty-five-page novella that most closely mirrored the future novel.

Unlike Humbert Humbert, the narrator of *Volshebnik* is nameless (though Nabokov once referred to him as "Arthur"). He does not have Humbert's artful insolence. Instead he is in torment from the first sentence, "How can I come to terms with myself?" A jeweler by trade, he moves back and forth between being open about his attraction to underage girls and his resolve to do nothing about it, coupling his inner torment to overweening self-justification. "I'm no ravisher," he declares. "I am a pickpocket, not a burglar." Humbert would sneer at the hypocrisy of this declaration.

Nabokov was not the artist he would later become, and it shows in the prose: "I'm not attracted to every schoolgirl that comes along, far from it—how many one sees, on a gray morning street that are husky, or skinny, or have a necklace of pimples or wear spectacles—*those* kinds interest me as little, in the amorous sense, as a lumpy female acquaintance might interest someone else." He doesn't have the wherewithal to describe his chosen prey, whom he first sees roller-skating in a park, as a nymphet. Such a word isn't in his vocabulary because it wasn't yet in Nabokov's.

Still there are glimpses of *Lolita*'s formidable style, as when *Volshebnik*'s narrator comments on "the radiance of [one girl's] large, slightly vacuous eyes, somehow suggesting translucent gooseberries" or "the summery tint of her bare arms with the sleek little foxlike hairs running along the forearms." Not quite up to the level and the hypnotic rhythm of Humbert's rhapsodizing about Dolores Haze ("The soot-black lashes of her pale-gray vacant eyes . . . I might say her hair is auburn, and her lips as red as licked candy"), but the disquiet is present, waiting to spring like a trapdoor.

As in the later novel, Nabokov's narrator preys upon his underage quarry through her mother. She is more broadly cast than Charlotte Haze, whose rages against and aspirations for her daughter make her an interesting figure. The mother here is little more than a cipher, a plot device to engineer the man and girl toward their fates.

Volshebnik's narrator may be tormented by his unnatural tastes, but he knows he is about to entice his chosen girl to cross a chasm that cannot be uncrossed. Namely, she is innocent now, but she won't be after he has his way with her. Humbert Humbert would never be so obvious. He has the "fancy prose style" at his disposal to couch or deflect his intentions. So when he does state the obvious—as he will, again and again— the reader is essentially magicked into believing Dolores is as much the pursuer as the pursuee.

Both men's plans are the same: "He knew he would make no attempt on her virginity in the tightest and pinkest sense of the term until the evolution of their caresses had ascended a certain invisible step," says *Volshebnik*'s narrator. He also sets the same stage for his seduction, in a faraway hotel, away from knowing, prying eyes, or so he thinks. The hotel, in Europe, is less shabby than Humbert's choice of The Enchanted Hunters, but serves the same purpose: allowing the narrator to watch over the sleeping girl and make his move against her will.

The outcome differs from *Lolita*. The narrator is consumed by the girl lying supine on the bed, robe half-open, and begins "little by little to cast his spell . . . passing his magic wand above her body," measuring her "with an enchanted yardstick." Here, again, Humbert Humbert would sneer. But then he did not have the girl look "wild-eyed at his rearing nudity," caught out like the pedophile he is. Nor does Humbert become "deafened by his own horror" when the girl begins to scream at his rejected advances. Humbert is all about self-justification; *Volshebnik*'s narrator suffers no such delusion about his quarry.

He tries to soothe the girl—"be quiet, it's nothing bad, it's just a kind of game, it happens sometimes, just be quiet"—but she will not be placated. And when two old women burst into the room, he flees, only to be hit by a truck, the ensuing gory mess described as "an instantaneous cinema of dismemberment." The narrator's fate is awful and inevitable. He is the predator hunted, captured, taken down. The girl's big bad wolf is punished by a passing truck.

Nabokov did not publish *Volshebnik* during his lifetime because he knew, as was clear to me upon reading it, that the story was not a stand-alone work but source material. It is more straightforward and less sophisticated than *Lolita*. As the scholar Simon Karlinsky wrote when *Volshebnik* was finally published in English as *The Enchanter* in 1986, the novella's pleasure is "comparable to the one afforded by studying Beethoven's published sketchbooks: seeing the murky and unpromising material out of which the writer and the composer were later able to fashion an incandescent masterpiece."

In other words, the story carries equal value to the creation of *Lolita* as did the story of Sally Horner. One was fiction; the other was truth. But art is fickle and merciless, as Nabokov explained repeatedly throughout his life. *Volshebnik* possesses a powerful engine of its own. It does not possess *Lolita*'s literary trickery and mastery of obfuscation, which continue to make moral mincemeat out of the novel's wider readership. Here, instead, is a more prosaic depiction of deviant compulsion and tragic consequences.

TWO OTHER WORKS are notable influences upon *Lolita*. Annabel Leigh, Humbert Humbert's first love, is named in homage to Edgar Allan Poe's poem "Annabel Lee." The novel's working title, *The Kingdom by the Sea*, is a quote from that poem, and Humbert's memories of his Annabel, dead of typhus four months after their seaside near-consummation, echo many more of Poe's lines. (Nabokov: "I was a child and she was a child." Poe: "I was a child and she was a child.")

Lolita also owes a great deal to an influence never explicitly referenced in the text, but one Nabokov knew well from translating into Russian in his early twenties: *Alice's Adventures in Wonderland*, by Lewis Carroll. As he later explained to the literary critic Alfred Appel:

"[Carroll] has a pathetic affinity with Humbert Humbert but some odd scruple prevented me from alluding in *Lolita* to his wretched perversion and to those ambiguous photographs he took in dim rooms. He got away with it, as so many Victorians got away with pederasty and nympholepsy. His were sad scrawny little nymphets, bedraggled and half-undressed, or rather semi-undraped, as if participating in some dusty and dreadful charade."

Perhaps a similar "odd scruple" may explain why Nabokov was quick to deny any connection between *Lolita* and a real-life figure he knew early on in his American tenure. Henry Lanz was a Stanford professor of motley European stock, "of Finnish descent, son of a naturalized American father, born in Moscow and educated there and in Germany." He was fluent in many languages, an avid chess player. By World War I Lanz was in London, married, at the age of thirty, to a fourteen-year-old.

Not long after the Nabokovs immigrated to America in May 1940, arriving in New York on the SS *Champlain*, Lanz arranged for Nabokov to teach at Stanford. Their friendship grew over regular chess games; Nabokov beat Lanz more than two hundred times. Over these jousts Lanz revealed his predilections—specifically, that he most enjoyed seducing young girls and he loved to watch them urinate. Four years later, Lanz was dead of a heart attack at the age of fifty-nine.

Nabokov's first biographer, Andrew Field, suggested that Lanz was a prototype for Humbert Humbert. Nabokov, however, denied it: "No, no, no. I may have had [Lanz] in the back of my mind. He himself was what is called a fountainist, like Bloom in *Ulysses*. First of all, this is the commonest thing. In Swiss papers they always call them un triste individuel."

Such a denial makes sense, in light of other future denials of real-life influence. Yet the months Nabokov spent being peppered with stories from a known pederast could not help but inform his fiction—and further bolster his involuntary, unconscious need to unspool *this* particular, horrible narrative.

SEVEN

Frank, in Shadow

Unlike Humbert Humbert, there was nothing erudite about Frank La Salle. His prison writings are unreliable, lacking the silky sheen that is *Lolita*'s narrative hallmark; grammatical mistakes pepper La Salle's rambling and incoherent oral and typewritten declamations. When he was employed, irregularly at best, he worked blue-collar jobs, a far cry from teaching foreign languages.

La Salle was a crude, slippery figure, who lied so much in middle age that it was difficult, at times impossible, for me to verify the facts of the first four decades of his life. One pseudonym dead-ended into another. Calls and emails to helpful, friendly archivists around the country bore no fruit, save for sympathy, until a recently discovered trove of documents filled in some lingering gaps.

I still don't know the substance of La Salle's childhood

and upbringing, and how early his predilections asserted themselves, making it difficult for me to determine where he came by his long-running desire for young girls. La Salle behaved as a pedophile, but it's hard to say whether that was his orientation—compulsion spurring opportunity—or impulsively seizing upon opportunity as a means of asserting power. Whatever he was dwarfed what he did.

A likely birth date is May 27, 1895, give or take a year. Frank La Salle was probably not his birth name, but it was almost certainly one of French-Canadian origin. Once, he said that his parents were Frank Patterson and Nora LaPlante. Another time he wrote down their names as Frank La Salle and Nora Johnson. He served a four-year prison sentence at the Leavenworth, Kansas, federal prison between 1924 and 1928 for running a car theft ring*. He hailed from Indianapolis, or Chicago, or maybe from Montreal. He needed a new origin story every time he changed aliases, at least twenty by my count, among them Patterson, Johnson, LaPlante, Campbell, and O'Keefe. As far as I could determine, the first name never varied.

For someone who shrouded his life in secrecy, it seems fitting that one of his most notorious aliases was that of Frank Fogg.

It is as Fogg that a sharper picture forms of the man later known as La Salle. In the summer of 1937, Fogg had a wife and a nine-year-old son. They lived in a trailer in Maple Shade, New Jersey. He claimed that his wife took their son and ran away

---

* For more on La Salle's early life and prison time at Leavenworth, see the Afterword.

with a mechanic. It's possible that might be true. By July 14 they were gone, and just over a week later Fogg himself would become a fugitive, with a new wife in tow.

He met her at a carnival: Dorothy Dare, not quite eighteen, with brown curly hair that framed an openhearted, bespectacled face. Born in Philadelphia, the oldest of six, Dorothy lived with her family in Merchantville, a ten-minute car trip from Maple Shade, and had graduated high school just the month before. Fights with her father over his strict parenting had grown so tense that Dorothy looked for every chance to escape. At the carnival, she found it in the man calling himself Frank Fogg.

He was more than twice as old as Dorothy, but she didn't mind the age difference. He wanted to marry her and she thought it was a terrific idea to elope. Which they did, only a few days after meeting, to Elkton, Maryland, the "Gretna Green of the United States," where weddings happened fast with few questions asked.

Dorothy's father, David Dare, was livid. Though they fought, he knew Dorothy was fundamentally a good girl. Even if she was not, technically, a minor, she was young, and this Fogg fellow was clearly not. When Dare discovered that Fogg was using a fake name, and was actually married, he got local police to swear out an eight-state warrant for the man's arrest on kidnapping and statutory rape charges on July 22, 1937. He claimed that Dorothy was fifteen, and thus a minor. The law caught up with the couple ten days later.

Cops arrested La Salle, still using the Fogg alias, in Roxborough, Pennsylvania, where he'd found a job, and took him to jail in Haddonfield, New Jersey. The charge: enticing a minor. Bail: withheld. Police simultaneously picked up Dorothy in the Philadelphia neighborhood of Wissahickon, where the couple had rented a room, and also brought her to jail. The two had a surprise for the arresting officers: Dorothy was not a minor, their Elkton marriage was legit, and Frank had the certificate, dated July 31, to prove it.

"He told me the truth," Dorothy cried, nervously fingering the shiny gold ring on her left hand. "I know he did. He couldn't have been married before. But if he did—oh, I'd just want to die!" Not long after uttering those words, Dorothy was released from jail, and slipped away from her parents, not yet ready to give up on her new husband, Frank.

The next morning, La Salle appeared in Delaware Township court. Dorothy was not there; nor did anyone know her whereabouts. Her father, however, was very much present. When he spotted La Salle, he punched the other man in the jaw. Dare grew even more furious when the presiding judge, Ralph King, dismissed the charges against La Salle, after the man testified that Dorothy had gone with him of her own will, and that they were lawfully married.

"I'll lock you up if you aren't careful," King warned Dare after he raised his voice in court one too many times demanding La Salle be held. But in the end, Dare got his wish, because the court wasn't done with Frank.

A day after the eight-state warrant had gone out on the wire, there was a hit-and-run accident near Marlton. A car resembling the one La Salle drove collided with a car owned by a man named Curt Scheffler. The driver of the first car fled the scene. La Salle, in court, denied he had been the driver. Justice of the Peace Oliver Bowen disagreed. On August 11, 1937, La Salle was fined fifty dollars and sentenced to fifteen days in jail. He also received an additional thirty days' sentence after failing to pay a two-hundred-dollar fine for giving false information. When he got out of jail, Dorothy was waiting. They picked up their marriage where it had been interrupted, and, apparently, the next few years were happy ones.

Dorothy and Frank, who cast off the Fogg alias and was La Salle once more, moved to Atlantic City. Their daughter, Madeline (not her real name), was born in 1939, and the young family were living at 203 Pacific Avenue when the Census came knocking a year later. So, too, did police, who this time arrested La Salle on bigamy charges. Few details are available—was it the earlier wife or a different woman?—save that La Salle wriggled out of it with an acquittal.

Two years later, when Madeline was three, Dorothy sued Frank for desertion and nonpayment of child support. Dare family lore had it that Dorothy discovered her husband in a car with another woman, and grew so enraged she hit the other woman over the head with her shoe.

What was passed down as a dark but amusing family story turned out to hide a more sinister truth. What Dorothy Dare discovered about her husband first came to light in the wee hours of March 10, 1942.

THREE CAMDEN POLICE OFFICERS walked into a restaurant on Broadway near the corner of Penn and spotted a girl sitting alone in a booth. Women sitting by themselves in public at three in the morning still stand out. Imagine what the cops thought in the early 1940s when they stumbled across a twelve-year-old girl all on her own so late in the night.

When acting sergeant Edward Shapiro and patrolmen Thomas Carroll and Donald Watson asked the girl what she was up to, "being out alone at such an hour," she evaded their questioning. So the policemen took her back to headquarters, where a city detective would ask the questions.

Under gentle coaxing by police sergeant John V. Wilkie, the girl opened up. She admitted she'd been out that night because she "had a date with a man about 40 years old." The man's name, she said, was Frank La Salle. He'd given her a card with the phone number and address of the Philadelphia auto body shop where he worked.

In his report, Wilkie wrote that the girl said La Salle had "forced her into intimacies." The girl almost certainly used plainer language. She also told Wilkie that La Salle made her introduce him to four of her friends by threatening to tell her mother what she had done with him.

The five girls were Loretta, Margaret, Sarah, Erma, and Virginia.* From the available records, it's not clear which of them was the one in the diner, but based on their birth dates, it was likely Loretta or Margaret. (Sarah, the oldest, had just turned fifteen.) All of them lived in Camden County, either in the city or in nearby Pennsauken. All of them were named in a 1944 divorce petition by Dorothy Dare as having "committed adultery" with her husband.

When police brought the other girls in to be questioned, Wilkie reported, each of them also told of "how they had been raped by La Salle."

SERGEANT WILKIE SWORE OUT a warrant for Frank La Salle's arrest, alerting police in Philadelphia of the twelve-year-old girl's sickening allegation. But when police showed up at La Salle's workplace, he wasn't there. They didn't find him at his

---

* I am withholding their last names to protect the privacy of their families, and because of the difficulty in locating descendants to verify the details.

last known address, either. Who knows how La Salle learned the police were coming for him, but he had fled. What's more, police learned, he'd gone back to his earlier alias of Fogg. They dug up an address in Maple Shade, then received word that he, Dorothy, and Madeline had moved back to Camden.

Police got a tip La Salle and his family now lived at a house on the 1000 block on Cooper Street. They kept the place under constant surveillance, hoping he might turn up. On the evening of March 15, a car pulled up in front of the house. The car had a license number linked to La Salle.

Detectives rushed into the house. They found and arrested a nineteen-year-old man who claimed he was La Salle's brother-in-law. But no La Salle. "We found out later," Wilkie said, "that as detectives walked up the front steps, La Salle made his escape out the back door."

For nearly a year, La Salle eluded the law. An official indictment for the statutory rape of the five girls came down on September 4, 1942. Tips streamed into Camden and Philadelphia police placing him in New Jersey, and sometimes in Pennsylvania, but nothing panned out—not until the beginning of February 1943, when cops got a tip that La Salle now lived at 1414 Euclid Avenue in Philadelphia, in the heart of where Temple University stands today.

On February 2, police descended upon the house and found La Salle, alone. They arrested him, taking him back to Camden to be arraigned. The Camden city court judge who signed the indictment and oversaw the February 10 hearing was a man named Mitchell Cohen. The two men would meet again, seven years later, in even more explosive circumstances.

La Salle pleaded not guilty to the multiple rape indictments from the Camden grand jury, but on March 22, 1943, he changed his plea to *non vult*, or no contest. The presiding

*Mug shot of Frank La Salle taken upon the start of his prison
sentence for the statutory rape of five girls, 1943.*

judge, Bartholomew Sheehan, sentenced La Salle to two and
a half years on each rape charge, to be served concurrently at
Trenton State Prison.

WHILE LA SALLE was incarcerated, Dorothy and Madeline had
moved back to Merchantville to be closer to Dorothy's parents.
She moved quickly to divorce him, filing a petition on January 11, 1944, stating that La Salle had "committed adultery"
with the five girls beginning on March 9, 1942—the night the
first girl reported her rape to police—and "at various times"
between that date and February 1943, when La Salle was finally
arrested. Frank wrote her frequently from prison—a habit he
would repeat later in life—but if he meant to persuade Dorothy
to stay married to him, he was not successful.

La Salle was paroled on June 18, 1944, after fourteen
months in prison. He took a room at the YMCA on Broadway
and Federal, registered for the draft, and got his Social Security
card. He also had to register with the city as a convicted crimi-

nal; a blurry photo from June 29, 1944, shows a middle-aged man with gray hair, blue eyes, high cheekbones, and a squint. He wore a more subdued expression than in the prison intake photo from March 1943, where he'd smirked at the camera, seemingly free from worry or care.

La Salle found work as a car mechanic in Philadelphia, but found himself in repeated trouble with the law. An indecent assault charge was dropped on Halloween, but the following August, he got caught at Camden's Third National Bank trying to pass off a forged $110 check. He was indicted the following month and swiftly convicted for "obtaining money under false pretenses." His divorce from Dorothy also moved toward completion that same month. The family court judge awarded full custody of Madeline to her mother on August 21. The divorce was final on November 23.

La Salle returned to Trenton State Prison on March 18, 1946, to serve eighteen months to five years on the new charges. The clock also began again on the balance of the statutory rape sentence. La Salle finished up both those sentences in January 1948, and was paroled again on the fifteenth of that month.

Now that he was back on the streets, it seems likely La Salle went to the downtown Camden YMCA for a cheap place to stay. It was across the street from Woolworth's, where weeks later, on a crisp March afternoon, he would spy a ten-year-old girl attempting to steal a five-cent notebook.

EIGHT

"A Lonely Mother Waits"

The more months that slipped by without answers, the greater the existential toll on Sally Horner's family. Ella bore the brunt of it. Sally was her daughter. She'd let the girl go off with a stranger because he'd said he was the father of school friends, who were waiting for her down by the Jersey Shore. She believed the lies the man forced her daughter to tell, and now the girl was gone.

Perhaps Ella entertained fleeting thoughts that Sally was dead, but she never admitted it in public. She'd found work as a seamstress for Quartermaster Depot, so at least there was enough money coming in to keep the lights on in the house, as well as the telephone in operation. Sally hadn't called home again, but if she ever chose to, it would be a disaster if the line was disconnected.

Ella aired her anguish in a December 10, 1948, article published by the *Philadelphia Inquirer,* headlined "A Christmas Tree Glows, a Lonely Mother Waits." Sally had been missing for nearly half a year by that point. Ella kept a figurative candle burning for the time when her daughter would return home safe. And with Christmas a little more than two weeks away, the tree Ella set up was, according to the unbylined reporter, "freighted with memories of other and happier Christmases," a means of multiplying the "candle's tiny gleam." Ella could think of no better way to express her faith that Sally would come back to her.

When that happened—Ella could not allow herself to think "if," only "when"—Sally would have to contend with one very big change to the family. Her niece, Diana, was five months old when the article appeared in the *Inquirer.* But Diana's arrival couldn't help but be bittersweet for her grandmother, Susan, and her husband, as long as they had no clue to Sally's whereabouts. Sally had so looked forward to being an aunt.

The young couple buried their uncertainty and despair in the daily care of their new baby. Changing diapers, rocking her to sleep, trying to get much-needed rest in small bursts. They also had the family greenhouse to look after—the flowers and plants wouldn't grow themselves.

Ella was thrilled to be a grandmother. But her joy was always tempered so long as Sally wasn't here with her. And as she told the *Inquirer,* the moment Sally walked back through the front door at 944 Linden Street, she would not be punished in any way. "Whatever she has done, I can forgive her for it. If I can just have her back again."

SALLY HORNER'S TWELFTH BIRTHDAY, on April 18, 1949, came and went with no news. Had she vanished into the ether?

Would her body turn up? Or was Sally out there, ready to be found, hoping that she would come home again? Camden police kept the case open, and Marshall Thompson tracked every lead.

The case had taken on added urgency a month earlier, on March 17, when the Camden County prosecutor's office added a second, more serious, indictment of kidnapping to the existing charge against La Salle. Where abduction carried a maximum prison sentence of only a few years, a kidnapping conviction upped the ante to between thirty and thirty-five years—effectively, for someone of Frank La Salle's age, a life sentence.

No documents have survived to explain the more serious charge. Perhaps the prosecutor, or the Camden Police Department, had received a credible tip to La Salle's whereabouts and hoped that news of the new indictment might flush him out. Or perhaps there were concerns about the statute of limitations on the original abduction charge if Sally stayed missing for a long time, or even forever.

The media moved on from covering the story. No one marked the first anniversary of her disappearance. At Christmas 1949, they did not publicize appeals for Sally's safe return. Other sensational local crimes pushed her out of the papers, including a mass shooting on Camden's East Side and another mysterious disappearance: that of the wife of Jules Forstein, a Philadelphia magistrate.

Dorothy Forstein's disappearance baffled investigators. She had spent the previous four years in a state of anxiety after an unknown assailant nearly killed her just outside her front door. On the evening of October 18, 1949, Dorothy's husband, Jules, attended a party on his own. He said he asked Dorothy to come with him, but she insisted she'd rather stay home with the children, stepdaughter Marcy, nine, and their seven-year-old, Edward.

When Jules got home at 11:30 P.M. Dorothy was gone and the children were frantic. They claimed a stranger had come into the house, knocked their mother unconscious, and then hefted the five-foot-two, 125-pound woman, clad in pajamas and red slippers, over his shoulder, carrying her out and locking the door behind them. Before he left, he patted Marcy on the head and told her, "Go back to sleep." Marcy recalled that he was wearing a brown cap and "something brown in his shirt."

Jules said he did not report Dorothy missing right away because he thought Marcy was telling a fib, and believed—oddly, considering her near-agoraphobia—that his wife must still be in the neighborhood. He called police four and a half hours later, just before 4:00 A.M. The police also brushed off Marcy's story as fantasy. Then a female psychiatrist was called in to question the little girl, and after several lengthy conversations, the psychiatrist concluded Marcy was telling the truth about what happened that night.

Over the next few days, people reported sightings of the missing woman all over the Philadelphia area. Camden also figured in the initial investigation. The Friday night after Dorothy vanished, Camden patrolman Edward Shapiro noticed a blond woman hovering around the corner of Broadway and Fourth Street. Shapiro told Philadelphia detectives he first saw the woman coming out of a telephone booth next to a candy store. She seemed startled to see him, and he was himself startled by how much she resembled Dorothy Forstein. Shapiro followed her to a tavern, where she ordered a beer, but when the woman noticed him again, she split.

The following night, Shapiro saw the woman again on the same corner, and overheard her speaking with a male companion. "I've only got one arm," he heard the woman say. This statement caused detectives to perk up. One of Dorothy's lingering injuries from her earlier attack was a recurrent dislocated shoulder that landed her in the hospital for treatment several times a year. Detectives were certain Dorothy, even under a false name, might turn up at a hospital to have her shoulder reset again. She did not. A thousand-dollar reward offered by Jules Forstein yielded no new leads. Dorothy was declared legally dead eight years later, in 1957, a year after her husband had died of a heart attack at home.

SALLY'S FAMILY took the waning public interest as a sign that it was better not to speak of her disappearance, even among themselves. Her absence was a low thrum, ever-present but unacknowledged. Of course they worried. Of course they feared for her safety. But no news meant no answers, and her fate was beyond their control. It was better to carry on with life.

Baby Diana turned one in August 1949. She was, by her parents' account and her own later recollection, a happy little girl, eager and talkative, who loved eating Grape-Nuts and drinking apple juice. Al ran the greenhouse, and Susan stopped in when she was up to it. Ella remained at 944 Linden Street, but living there was like a nightmare. "It was so different when Sally was here," she later recalled. "She was so cheerful and full of life."

Ella had difficulty sleeping. Many times in the night she would leave her room and go to Sally's. She would take out her daughter's toys and games and "just sit there and look at them." Ella washed and rewashed Sally's clothes "so they would be ready for her when she came back."

As the months wore on, she lost jobs and found others.

Sometimes she couldn't pay her bills for weeks. Sometimes the phone got disconnected, or the electricity was shut off. Periodically, Ella would go to Florence to stay with Susan and look after her granddaughter. Otherwise, Ella was alone. Alone to contemplate again and again the ways in which she held herself responsible for Sally's disappearance.

NINE

The Prosecutor

Sally Horner disappeared a few months after Mitchell Cohen was appointed as prosecutor for Camden County, a ten-year term that would last until 1958. He was already on the way to becoming the pivotal law enforcement figure in the city, a status that saw his name emblazoned, decades later, on the downtown federal courthouse. Taking on the prosecutor gig only bolstered his reputation. In the late 1940s, Camden County did not have enough major crime to justify a full-time prosecutor. So Cohen worked in spurts, spending the rest of his time on Republican party politics. He was so successful at it that he became the state party's de facto leader, allied closely with the New Jersey governor of the day, Alfred Driscoll.

Cohen's term as prosecutor was one of many jobs he held in law enforcement over a long legal career that stretched from early private practice with local law firms all the way to chief judge for the federal district court of New Jersey. He moved smoothly between prosecuting criminals and delivering judgments. Cohen wasn't one to bask in career glories, though. He was far too busy working to spend much time reflecting on the past.

Cohen did, however, take the time to dress the part of a big-shot lawyer. In his bespoke suits, he cut a figure that landed somewhere between David Niven and Fred Astaire. One lawyer told Cohen's son, Fred, "Whenever I appeared in front of your father, I felt I wanted to wear a white tie and tails and be at the top of my game because he was a classy guy."

Any resemblance Cohen bore to famous actors was, perhaps, intentional. He caught the bug for theater early in life, making trips up to New York to see Broadway shows a priority. In his younger years, Cohen spent his Saturday nights in line for concerts and stage performances at Philadelphia's Academy of Music, angling for a twenty-five-cent seat in the gallery. Once established in his legal career, he could afford to split a box seat with a friend.

He'd met Herman Levin at South Philadelphia High; they remained pals even after Cohen's family moved to Camden just before his senior year. The boys made a point of seeing every play in Philadelphia—long a city where Broadway-bound shows worked out problems and test-ran productions on audiences—on opening night. Levin ended up producing musicals like *Gentlemen Prefer Blondes, Destry Rides Again*, and *My Fair Lady*. Before the latter show opened in 1956, Levin told Cohen to "hock everything he had in the world" to invest in the production. Cohen did, and reaped the benefits as *My Fair Lady* smashed box-office records. Cohen also became a theatrical producer himself, cochairing the short-lived Camden County Music Circus during the summers of 1956 and 1957.

But Cohen's fashion sense, his theatrical interests, and even his political machinations did not overshadow his commitment to jurisprudence. Cohen cared about the law and about being fair. He believed it was as important to know when *not* to prosecute a case as when to prosecute. In 1938, early in his tenure as acting judge for the city of Camden, a husband and wife appeared together in his courtroom after she had attempted suicide by poison—back then a punishable crime.

"We had a quarrel, and I thought he didn't love me anymore," the twenty-eight-year-old woman told Cohen.

"Do you?" Cohen asked her husband, who was twenty-nine.

"I sure do."

"Then go home and forget about it."

Cohen didn't seek out notorious cases. They found him. When he won a trial, he didn't dwell upon the details. Those cases, and how Cohen approached them, are an important window into 1940s Camden, as well as the forces that set the city up for great societal changes.

WHEN MITCHELL COHEN set out to prosecute the men responsible for the murder of Wanda Dworecki in the fall of 1939, he had never worked on a capital case before, even though he'd been appointed city prosecutor for Camden three years earlier. Murders were rare then, a far cry from the statistics that designated the city as America's murder capital as recently as 2012. The strangulation of a girl a month shy of her eighteenth birthday stood out in its singular brutality.

Wanda's body was discovered on the morning of August 8, 1939, in an area near Camden High School frequently used as a lovers' lane. A corsage of red and white roses adorned her neck. The killer strangled her so forcefully that he broke her collarbone and breastbone. Then he dropped a rock onto her head, fracturing her skull.

Police working on the case weren't all that surprised Wanda had died so violently. Four months before her murder, in April 1939, two men had accosted Wanda on the street and thrown her into their car. They beat her up—a near-fatal assault—and tossed her out into a field in a desolate part of Salem County, just south of Camden. She spent weeks recovering in the hospital.

That beating wasn't Wanda's first brush with violence. In late 1938, she and a friend had been out walking in the neighborhood when several men tried to kidnap them. Police were convinced—or at least, this is what they said—that Wanda was "destined to be murdered." What they would soon learn is how much one man manipulated things so that destiny would become reality.

Wanda's father, Walter Dworecki, who had emigrated from Poland in 1913, appeared to be an upstanding figure. He preached at the First Polish Baptist Church, a congregation he'd founded when the family moved to Camden from rural Pennsylvania. His teenage daughter troubled him, especially after her mother, Theresa, collapsed at the breakfast table and died in 1938. He brooded over Wanda's fondness for the opposite sex, lecturing her about preserving her virtue, verbally abusing her in such a way to make an example of her to her younger siblings, Mildred and Alfred.

Respectable appearances can hide awful secrets, and Dworecki had plenty. Like being out on bail for setting fire to a house in Chester, Pennsylvania, in a scheme to collect insurance. Or having been sentenced to five years' probation for passing counterfeit money. Like getting so angry at a neighborhood boy that he allegedly fractured the teen's jaw with a broomstick. Or taking out a $2,500 life insurance policy (nearly

$45,000 in 2018 dollars) on his wife, Theresa, whose cause of death was officially "lobar pneumonia"—the same cause of death listed for a number of victims of a murder-for-insurance scheme in Philadelphia whose culprits had some connections to the preacher.

The secret Walter Dworecki should have tried harder to keep was his fondness for hanging around Philadelphia dives, looking for men who might be willing and able to kill his daughter.

Immediately after Wanda's murder, Dworecki slipped into the role of grieving father. He cried, "My poor Wanda!" when he saw his daughter's body at the morgue, and then fainted. He had an alibi for the time of her death, but his grief-stricken act cracked quickly once police started to investigate.

A witness had seen Wanda with a "large blond gentleman" the night before she was murdered, who turned out to be twenty-year-old Peter Shewchuk, who boarded at the Dworeckis' and romanced Wanda every now and then. When Shewchuk learned he was wanted for questioning, he fled Camden for his boyhood home in rural Pennsylvania. Police caught up to him on August 27, after his father turned him in.

In the interview room, all it took was the offer of a single cigarette for Shewchuk to open up. He told the detectives that he and Dworecki had met up in Philadelphia earlier in the evening of August 7. "He gave me 50 cents to cover my expenses and then went to conduct his religious services. I met Wanda and we strolled down the street." Walking past the lovers' lane, Shewchuk said he "suddenly felt the urge to kill" Wanda the way her father had told him to: "Choke her, hit her with a rock, twist her neck." Dworecki was supposed to pay Shewchuk a hundred dollars for the murder, but after Wanda was dead, he reneged on the deal.

Armed with Shewchuk's confession, Camden police brought in the preacher. It turned out Dworecki had taken out a life insurance policy on Wanda around the same time as he had taken one out on her mother. He had hired three men, including Shewchuk, to kidnap and kill Wanda back in April. When she survived the failed attempt, Dworecki upped the policy on his daughter to nearly $2,700, with a double indemnity clause should she die in an accident, and got ready to try again.

Dworecki eventually confessed in a statement that ran to nearly thirty pages. The preacher admitted to being aggrieved by Wanda's behavior, but he claimed the idea to kill Wanda originated with two men, Joe Rock and John Popolo, whom he'd met in Philadelphia. Dworecki said they urged him to kill his daughter to collect the insurance money and pressed him again as time went on. He roped Shewchuk into the murder plot after learning the younger man had bragged about sleeping with Wanda. Shewchuk denied it, but Dworecki sensed an opportunity to manipulate the boy into carrying out his murderous scheme.

Both men entered guilty pleas in Camden County Court on August 29, 1939. Dworecki refused to look at anyone. Shewchuk chose the opposite tack, smiling whenever someone caught his eye. Then Mitchell Cohen, the city prosecutor, explained to the assembled crowd, including the surprised defendants and their lawyers, why the guilty pleas had to be thrown out. New Jersey state law at the time stipulated that one was not allowed to be sentenced to death if he pleaded guilty. Capital cases, and murder certainly counted, required that the defendants face a full trial, verdict, and sentencing.

Cohen voided the pleas, then bound the cases over to the county court, where Samuel Orlando (whom Cohen would succeed a few years later) would prosecute them. Cohen's work was done, but he paid attention to what happened in the county courtroom. Orlando cross-examined both Shewchuk and Dworecki with extra vigor. Shewchuk received a life sentence in exchange for being the primary witness against Dworecki. The preacher's confession was admitted into evidence, despite his lawyer protesting it should be kept out. The jury found him guilty after swift deliberations.

Shewchuk was paroled in 1959 after surviving his own near-fatal beating in prison; he died in the late 1980s. Dworecki was put to death by electric chair on March 28, 1940. Before his execution, he implored his surviving children, Mildred and Alfred, to lead pious lives and asked that "God have mercy on their souls." Dworecki's grave lies next to that of the daughter he murdered.

UNLIKE THE DWORECKI CASE, where Cohen did not have to work to establish the defendants' guilt, he played a larger role in a subsequent murder trial that garnered a great deal of media attention. This case concerned the death of twenty-three-year-old Margaret McDade (Rita to her friends) on August 14, 1945, as Philadelphia, Camden, and the entire country celebrated V-J Day. That night, Rita's best friend and fellow waitress Ann Rust saw her in the arms of a stranger, dancing to a Johnny Mercer tune. Five days later, Rita was found naked and dead at the bottom of a cistern near a sewage disposal plant. An autopsy determined that she had been raped, beaten bloody, and tossed into the cistern alive. She died of suffocation.

Not long after, police arrested the stranger Rita had danced with on the last night she was seen alive. Howard Auld was a former army paratrooper, recently discharged. When police found him, Auld gave them a fake name ("George Jackson") and claimed to be innocent. Discharge papers he carried caught him out on the first lie. Careful interrogation spurred Auld to confess to McDade's murder.

Auld recounted an all-too-familiar, all-too-horrible story: after the dance, he had made a move on Rita that she turned down. He got angry, punched her in the face, and choked her until she passed out. Auld claimed he felt for a pulse and, when he found none, dumped her into the cistern. (Never mind that she was still alive and he omitted mention of the rape.) Auld's time in the army also included repeated stints in a mental hospital and various bouts of violent behavior, which his lawyer, a court-appointed defense attorney named Rocco Palese, would use as a mitigating circumstance in the trial.

Auld was sentenced to death for Rita McDade's murder in 1946, but the conviction was tossed out on appeal several months later. The presiding judge, Bartholomew Sheehan, had failed to tell the jury that they could recommend mercy—meaning, a verdict other than death—in finding Auld guilty of first-degree murder. The Camden County prosecutor's office moved quickly to try Auld again, but proceedings did not begin until 1948, after Mitchell Cohen's appointment as top prosecutor.

Sheehan was also the judge for the second trial. Cohen asked for the death penalty, in accordance with New Jersey state law. Auld's new court-appointed attorney, John Morrissey—Palese had since been appointed as a judge—implored the jury to be lenient toward his client, "a feeble-minded boy," and deliver a not-guilty verdict by reason of insanity. But Cohen prevailed with the jury. Morrissey indicated he would appeal, and did, several times over, delaying the execution date a half dozen times. Howard Auld did not die in New Jersey's "Old Smokey" until March 27, 1951. His final words were "Jesus, have mercy on me."

BY THE END OF 1949, Mitchell Cohen had established his bona fides as Camden County prosecutor. He had tried one capital case directly and worked on another, even though he was deeply conflicted about the death penalty. Decades later his son, Fred, recalled Cohen becoming "very emotional" when the subject came up, so much so that they did not discuss it again. Cohen did his duty, whether asking for the harshest sentence as a prosecutor or delivering the sentence as a judge. But he did not have to like it and, with that single exception, took care not to bring his feelings home to the Rittenhouse Square town house he shared with his family.

He would also vault onto the national stage with his handling of a case that would shake the city to its foundation, and foreshadow similar massacres in the decades to come. But he did not close the books on Sally Horner's abduction. To his knowledge, the new kidnapping charge had not flushed out Frank La Salle. Sally was still missing. And the more time passed, the less likely the outcome would be a good one.

Baltimore

H ere's the point in the narrative where I would like to tell you everything that happened to Sally Horner after Frank La Salle spirited her away from Atlantic City to Baltimore, and the eight months they lived in the city, from August 1948 through April 1949. The trouble is, I didn't find out all that much. A scattershot list of addresses and court documents can't bring to life what a little girl thought or felt. Visiting the neighborhood where Sally lived, and walking by the school she attended, can't adequately bridge the decades. The neighborhood has changed, demographically and socio-economically. Sally, were she still alive today, would barely recognize it.

The meager paper trail frustrated me. My patience frayed as I ran up against dead end after dead end, record search after fruitless record search, to try to build up a picture of the months Sally lived in Baltimore. If she made friends, or had someone she felt she could trust, I couldn't find them. If there are people still living who knew her at the time, I could not track them down. If she kept a journal during her captivity, it did not survive. She did go to school in Baltimore—a Catholic school—but if any of its records remain, they are buried under decades of detritus no one has the inclination to sift through.

But I needed to understand what Sally was thinking and feeling—or at least approximate an understanding—so I read as many accounts as I could find by girls, born one or two generations after her, who survived years or decades of abuse by their kidnappers. I also examined kidnappings from the decade or so before Sally was taken.

Stranger abductions are rare now and were, perhaps, even rarer when Sally vanished. That's why the kidnapping of Charles Lindbergh, Jr., in 1932 caught America's attention and held it for weeks. The celebrity of the boy's parents, superstar pilot Charles Lindbergh and his wife, Anne Morrow Lindbergh, certainly helped, but the boy's snatching felt like the manifestation of every parent's worst fear—that their child might be stolen in the middle of the night from his bedroom by strangers—and kept the country gripped until the baby's body was discovered weeks later.

Abductions where the child is held for a significant period of time before being rescued alive occur with even lesser frequency. That's why, fourteen years before Sally Horner's abduction, the kidnapping of six-year-old June Robles, the daughter of a well-to-do Tucson, Arizona, family, stood out. A man driving a Ford sedan waited for June after school on April 25, 1934, and enticed her to get into his car. Several ransom notes arrived at the Robles household. The first demanded fifteen thousand dollars; the second, ten thousand. Days passed with false sightings and near-arrests, until a Chicago-postmarked letter delivered to Arizona governor B. B. Moeur's residence described where June was being held. A search in the Tucson desert turned up a metal box buried three feet underground. June, chained, malnourished, and covered in ant bites, was found alive inside.

For someone held captive in a tiny box for nineteen days, the girl was in remarkably good spirits. Several days after her rescue, June appeared at a press conference filmed by Pathé studios. (Reporters did not ask her questions, though, allowing her father, Fernando, to steer June through the session.) The little girl seemed poised, her answers sounding rehearsed. She said she was looking forward to going back to school that Friday. It was the last interview Robles ever gave. She never spoke to the media again.

As June's public silence stretched, so did the investigation. Leads proved false, no arrests were made, a grand jury failed to indict anyone, and the FBI eventually gave up, privately agreeing with the grand jury's conclusion of "alleged kidnapping." June stayed in Tucson, where she married and had children and grandchildren. By the time she died in 2014, in such obscurity that it took the press three years to connect her to her childhood ordeal, authorities still had no proper answer about who kidnapped her. It remains a mystery, as does the effect the kidnapping had upon June and her family.

Captivity narratives, such as the recent "found alive" stories of young women including Elizabeth Smart, Jaycee Dugard, Natascha Kampusch, and the trio Ariel Castro held prisoner in Cleveland, opened up a psychological trapdoor into Sally's probable state of mind. They also allowed me to understand how kidnappers were able to subject these girls and women to years of sexual, physical, and psychological abuse.

Smart, Dugard, and Colleen Stan—the "Girl in the Box" under her tormentors' sway for seven years—left their abductors' homes, shopped at supermarkets, and even traveled (Stan visited her parents while she was a captive) without asking anyone for help. They survived by adjusting their mental maps so that brutality could be endured, but never entirely accepted as normal. Every day, every hour, their kidnappers told these women that their families had forgotten all about them. Year after year, their only experience of "love" came from those who abused, raped, and tortured them, creating a cognitive dissonance impossible to escape.

Dugard's eighteen-year bond with her abductor resulted in her bearing two children by him. The fear of losing her daughters, no matter how squalid her situation, caused her to deny her real identity to the police at first, revealing the truth only when she felt secure that she was safe from her kidnappers. Smart, too, needed the same foundation of trust to tell law enforcement who she really was.

We know how these girls coped and felt because several of them published books about their extended ordeals. Smart, Dugard, and the Cleveland three—Amanda Berry and Gina DeJesus together, and Michelle Knight on her own—were able to tell their stories the way they wished and when they chose. In doing so they sought to make something meaningful of their lives.

Sally Horner did not have the chance to tell her story to the world, unlike the women and girls of later generations. She also didn't have the choice of keeping her account wholly private, unlike June Robles. What remains of her time on the road with Frank La Salle are bits and pieces cobbled together from court documents and corroborated by city records. Absence is as telling as substance. Inference will have to stand in for confidence. Imagination will have to fill in the rest.

THE SUMMER'S GREAT HEAT WAVE was some weeks away, but it still sweltered plenty on the Baltimore-bound bus. Frank La Salle and Sally Horner had taken a taxicab to the bus depot in Philadelphia. Perhaps Sally wondered why they were going so far out of the way if they were headed south. Maybe she asked why they had to leave Atlantic City so quickly, or where the station wagon had gone, or why they had to leave their clothes and photos behind. Most likely, she kept any complaints or questions to herself.

She had to keep remembering the script, that La Salle was her father. His word was law. She had to stick to the story to avoid punishment. She had to endure his daily torments. She had to retreat to her own mind to escape the void of her current situation.

The cab pulled up in front of the Philadelphia station. Frank and Sally made their way to the Greyhound bus bound for Baltimore before it pulled away at 11:00 A.M. He bought their tickets, Sally squeaking under the wire for the half-price fare. They settled in their seats for the three-hour trip. They may not have been alone. Sally later said that a woman she knew as "Miss Robinson" had joined them. La Salle had told her the woman was some sort of assistant or secretary. She was perhaps twenty-five, though an eleven-year-old girl's sense of how old people are can be skewed.

The Philadelphia Greyhound made one stop along the way, either in Wilmington or in Oxford, Delaware. After the short break, the bus moved over to Route 40, which turned into the

Pulaski Highway. Was Sally impressed by the wider lanes and speeding cars on the still-new highway? What did she allow herself to dream over the three-hour trip before the bus pulled into the downtown depot in Baltimore? Did she hope for a chance of escape, or had she resigned herself to being trapped by La Salle's new vision of her life?

They arrived in Baltimore just after 2:15 in the afternoon. "Miss Robinson," if she existed, vanished from the picture, perhaps as soon as they got off the bus, collected their luggage, and looked for a cab or local transit to take them to their lodgings. Most likely they ended up staying downtown that first evening and for the next few days, around West Franklin Street in the neighborhood of Mount Vernon. Blocks away lay the city's most prized landmarks, including City Hall, the Museum of Art, and the original Washington Monument. Testaments to Baltimore's beauty and power, but also a refuge out of Sally's grasp.

La Salle needed to find work right away. The Belvedere Hotel, a place so swanky that Woodrow Wilson, Theodore Roosevelt, and King Edward VIII and Wallis Simpson stayed there, may have hired him. It was less than a mile's walk from West Franklin Street. It would explain why La Salle listed a hotel bellman named Anthony Janney as a reference in later court documents. And what better place for a fugitive to hide than among hotel staff serving the toniest, richest guests in a Beaux Arts building nestled within Baltimore's most prominent neighborhood?

I was also struck, while walking around the district, by how close Sally was to the Enoch Pratt Free Library. It's a wonderful place for researchers, and a safe harbor for bookish types of all kinds. Sally loved to read; were books a way for her to imagine herself in different worlds she could control, or was the library yet another place she couldn't go, somewhere she fantasized about as a refuge from Frank La Salle's relentless assaults?

Because in Baltimore, something changed in their relationship. Publicly, they kept up the pose of father and daughter. In private, the power imbalance between them grew more noxious. It was in Baltimore, according to Sally, that rape became a regular occurrence. It was the place where Frank La Salle subjugated her totally to his will psychologically and physically. The outside world never had a clue, even after La Salle sent his "daughter" to school.

There was no way he could have kept her home if he wanted to maintain the illusion of normalcy. The summer was over and an eleven-year-old girl, shut away at home or loose on the streets while he was at work, would draw attention—and questions. La Salle couldn't control her every thought and move while she was at school, true. But by this point he'd broken her down enough, between the threats and the rapes, and the apologies and the treats, that he must have felt a measure of confidence that Sally would do exactly what he said, at all times.

To enroll Sally at Saint Ann's Catholic School, they had to leave West Franklin Street. So in September 1948, they moved to Barclay, a neighborhood on Baltimore's east side. There La Salle and Sally settled in an apartment on 437 East Twentieth Street, between Barclay and Greenmount Avenues, a block up from the local cemetery. At the time, the neighborhood was a middle-class enclave of brick town houses, where neighbors mingled freely if they wished, or kept to themselves if they did not. Over the next eight months, Sally got used to the new name Frank had given her: Madeline LaPlante.

Here's how I imagine Sally Horner's days during the 1948–1949 school year. She'd wake up, get dressed, act the part of daughter to her "daddy," and shove from her mind the fact that her current life was the opposite of normal. He probably took Sally to school for the first week, just to be sure she wouldn't do anything rash like speak out or run away. Afterward, he trusted Sally to go by herself. She knew he had to be at work early in a different part of town. She did not want to disappoint him. She resolved she never would.

She'd smile and nod to their landlady—Mary or Ann Troy; she got the two confused even though she'd been told over and over that they weren't related—and other neighbors as they headed off to work. Then, she'd walk west along East Twenti-eth Street. At the end of the block was the Diamond, the diner where she and La Salle took many of their meals, since he didn't have the time or the patience to cook, and she was still learning how. Sally usually skipped breakfast, waiting to eat until after morning prayers. Perhaps on some days, the waitress, Marie Farrell, packed up a piping-hot fried egg sandwich for her and put it on Frank's tab.

Breakfast in hand, Sally would turn right at the end of the block, walking up Greenmount Avenue until she reached the corner of Twenty-Second Street. There was Saint Ann's, an extension of a Roman Catholic church that had been in Barclay for more than a century. The schedule was strict. All students had to attend mass first thing in the morning. Sally sat with her classmates on uncomfortable pews as Monsignor Quinn, Saint Ann's pastor and principal, intoned the daily prayers in Latin and English. She kept an eagle eye out for Mother Superior Cornellous—the older woman did not tolerate her students fidgeting or misbehaving.

Then, if she had remembered to fast, Sally took Communion. The priest placed the host on her tongue. As it melted, Sally knelt and prayed for her eternal soul. Was the possibility of escape part of her prayers? Did she pray that someone would see behind the calm facade of Madeline LaPlante to the captive Sally Horner? Did she wonder if the things Frank asked her to do, which he said were "perfectly natural," were, in fact, a mortal sin? Or did she pray for things to stay as they were because they might get even worse?

When Communion ended, Sally went back to her pew. Mass was over, so it was time for the fried egg sandwich, now cool enough to eat, and then for her classes. So many hours in the day stretched ahead where all she had to think about was her studies. She had to do well and keep up her grades or else there would be more punishment at home, and so she likely did. But Sally also didn't want to draw undue attention to herself, in case someone—especially the Monsignor or the Mother Superior—grew suspicious and started asking too many questions. Better to embrace the invisibility. Better not to stand out.

When the last school bell rang and it was time to go home, Sally reversed her morning walk. But if there was time, or if she felt a smidgen bolder, perhaps she ventured up a block to Mund Park. The park was a place where the mind could roam and think of freedom. Where the green grass grew just like it did in Camden. Where she could think of her real home, and wonder if she would ever see it again.

I DON'T KNOW WHY La Salle chose to enroll Sally in Catholic schools, both in Baltimore and elsewhere. No one remembered him being a churchgoer or having any religious leanings. Before her abduction, Sally likely attended a Protestant church. One possible reason is that a parochial school did not have to conform to the same rules and regulations as public schools.

Catholic institutions were less likely to ask questions of a new student arriving later in the school year, under a false name, with dubious documentation at best. Instead of viewing a girl like Sally with suspicion, some opposite effect, like sympathy, may have prevailed.

But I suspect La Salle gravitated toward Catholic institutions because they were a good place to hide in plain sight. The Church, as we now know from decades' worth of scandal, hid generations of abused victims, and moved pedophile priests from parish to parish because covering up their crimes protected the Church's carefully crafted image. Perhaps La Salle saw parochial schools for what they were: a place for complicity and enabling to flourish. A place where no one would ask Sally Horner if something terrible was happening to her.

ELEVEN

Walks of Death

Back in Camden, Sally Horner's plight had been consigned to the same purgatory that befalls every long-
term missing child investigation. The city hadn't moved
on, but her fate was no longer the highest priority. Camden
residents wanted to embrace progress, to bask in fortunes they
believed would last forever. There was little warning of the out-
sized event that would bewilder them and foreshadow the pre-
cipitous decline in the city's near-future.

In the fall of 1949, Camden believed in its own prosperity.
It had weathered the Great Depression and near-bankruptcy
in 1936, the result of financial mismanagement by the local
government. Private industry still thrived. The New York Shipbuilding Corporation still had contracts from the navy and the
Maritime Administration. Smaller shipbuilding companies,
like John Mathis & Company, had doubled their workforce

during the Second World War and seemed primed to expand. Manufacturing jobs in the region were a year away from an all-time peak of 43,267. Campbell's Soup still employed thousands of workers at its local headquarters.

No company represented Camden's sense that the future was theirs for the taking more than RCA Victor, the phonograph company. In June 1949, it had introduced the "45," a smaller, faster alternative to Columbia's "LP" record format. RCA Victor also began producing technology for television, making equipment required by broadcast studios as well as for television sets regular home-buyers could acquire.

A great many forces underlay Camden's eventual negative transformation. But Sally Horner's abduction wasn't the spark. Rather, the morning of September 6, 1949, seems to me like the inflection point between progress and backlash, hope and despair, promise and decline. The scope of the crime seemed an unfathomable one-off, but its grotesque repetition in the decades to come demonstrates how a singular evil can become all too mundane.

AT EIGHT O'CLOCK that morning, a mother woke up her son for breakfast. He'd been out late the night before, sitting and stewing in a movie theater on Market Street in Philadelphia, waiting for a date who never showed. The son's homosexuality wasn't quite a secret, but nor could he flaunt it when sex between men was still very much against the law.

That his date, a man with whom he'd been in the midst of a weeks-long affair, stood him up was indignity enough. Then he'd returned to his home in Cramer Hill to find the fence he built to separate his house from his neighbors' home had been torn down.

The man drank a glass of milk and ate the fried eggs his mother, Freda, prepared. Then he went into the basement, whose walls were covered in memorabilia from the war he'd fought in, and where he had written down meticulous notes on each enemy soldier he'd killed. He regarded his nine-millimeter pistol, a Luger P08, for which he had two full clips and thirty-three loose cartridges, and thought about the list of people—neighbors, shopkeepers, even his mother—he wanted to wipe off the face of the earth.

He grabbed a wrench and went back to the kitchen. He raised it, threatening Freda. "What do you want to do that for, Howard?" she cried. When he didn't answer, she repeated the question as she backed away from him, then ran out of the house to a neighbor's. He retrieved his Luger and ammunition from the basement, as well as a six-inch knife and a five-inch pen-like weapon tricked up to hold six shells. Then he cut through the backyard and shot at the first person he saw: a bread deliveryman sitting in his truck.

Howard Unruh missed his first target, but he wouldn't miss many more. Twenty minutes. Thirteen dead. And a neighborhood, a city, and a nation forever marked by his "Walk of Death."

FOR MARSHALL THOMPSON, Unruh's murderous spree hit too close for comfort. He and his family lived around the corner from Unruh and his mother, at 943 North Thirty-Second Street. Most of those who died or were injured on the morning of September 6 were Unruh's neighbors on River Road, which was the main thoroughfare of East Camden.

Thompson might have gotten his hair cut at Clark Hoover's barbershop a few feet down River Road. That awful morning, the barber took a fatal shot, as did six-year-old Orris Smith, perched on a hobbyhorse inside the shop. If Thompson needed his shoes repaired and shined, he likely got it done at the repair shop next door, where Unruh killed the cobbler, John Pilarchik. Down the street was the tailor shop, owned by Thomas Zegrino. He was out when Unruh arrived, but Zegrino's new wife, Helen, was not, and she paid the price.

Unruh then shot Alvin Day, the television repairman. James Hutton, the insurance agent, made the dreadful mistake of running out of the drugstore to see what the commotion was all about, and also died. So, too, did a mother and daughter, Emma Matlack, sixty-six, and Helen Matlack Wilson, who'd driven in from Pennsauken for the day and failed to comprehend the massacre unfolding. Unruh shot them dead, and Helen's twelve-year-old son, John, took a bullet in the neck. He died the next day in the hospital.

Others were injured. Like Madeline Harrie, caught in the arm by a bullet after Unruh's first two missed, and her son, Armand, who tried vainly to tackle Unruh when he invaded their home.

Unruh moved on to his worst grudge late in the rampage, hunting down his next-door neighbor, Maurice Cohen, owner of the drugstore, to make him pay for the business with the fence as well as other perceived grievances. Not spotting him in the store, Unruh went upstairs to the family apartment. As Maurice climbed onto the roof, his wife, Rose, shoved their son, twelve-year-old Charles, into a closet, and then hid in a separate one. Unruh searched the apartment and then went out on the roof, where he caught a glimpse of Maurice running away. Unruh fired into the druggist's back. The shot jerked Maurice off the roof and he was dead before hitting the street.

Unruh went back inside and fired his Luger several times into the closet where Rose was hiding. She died instantly. Maurice's mother, Minnie, was in the bedroom, in a frantic state as she tried to get police on the phone, when Unruh caught up to her. He shot her in the head and body. She fell back on the bed and died there.

Charles stayed hidden until it was utterly quiet. When officers finally found him, he would not be comforted—he'd heard everything. Charles leaned halfway out his apartment window and screamed, "He's going to kill me. He's killing everybody."

Howard Unruh had walked down the stairs and made his way to the Harries' place. There he discovered he was out of ammunition. Hearing the police sirens, he doubled back to his mother's house to await his fate.

IN A LARGER CITY, where officers didn't walk the beats where they lived, Marshall Thompson might not have taken part in police efforts to apprehend Howard Unruh. But it's unlikely Thompson could have begged off even if he wished. One of Thompson's colleagues on the Camden detective squad, John Ferry, also lived in Unruh's Cramer Hill neighborhood. The summer before, Ferry had tried to help Unruh find a job as a favor to the man's uncle, a deputy fire chief.

Ferry had just finished up a midnight-to-eight shift. He was on his way home when he saw his insurance man dead in the street, as well as other victims. "When the other cops started arriving I went home and came back with my shotgun," Ferry recalled in 1974, the twenty-fifth anniversary of the massacre. With the body count rising and ambulances blaring to and from Cooper Hospital, Thompson was one of more than four dozen cops who descended upon Cramer Hill that morning.

Howard Unruh had barricaded himself in his home. It was up to a group of policemen led by Detective Russ Maurer to figure out how to coax him out. Maurer sidled up to the front of the house. A throng of cops, including Thompson, covered Maurer, poised to throw tear gas through the window if Unruh acted rashly. As *Courier-Post* columnist Charley Humes observed, "Russ [Maurer] could have paid with his life, because the killer seldom missed. That was a brave act."

John Ferry was crouched with several other policemen in Unruh's backyard, awaiting any sign of the man. When Unruh appeared in the window, Ferry turned to James Mulligan, his supervisor on the detective unit, and asked, "Jim, should I take his head off?"

"No," replied Mulligan. "There has been plenty of killing."

Unruh later told police he "could have killed Johnny Ferry . . . any time I wanted." Ferry's past attempt to find Unruh a job may well have saved his life, and perhaps the others. Unruh made his decision. "Okay. I give up, I'm coming down," he shouted to the cops down below.

"Where's that gun?" a sergeant yelled.

"It's on my desk, up here in the room," Unruh said, then repeated: "I'm coming down."

Unruh opened the back door and came out with his hands up. More than two dozen officers trained their guns upon him. One yelled, "What's the matter with you? You a psycho?"

"I'm no psycho," said Unruh. "I've got a good mind."

MITCHELL COHEN, THE Camden County prosecutor, had just returned from a summer vacation at the Jersey Shore. He expected his office would be its usual bustling self the morning after Labor Day, and that he would be faced with a fresh round

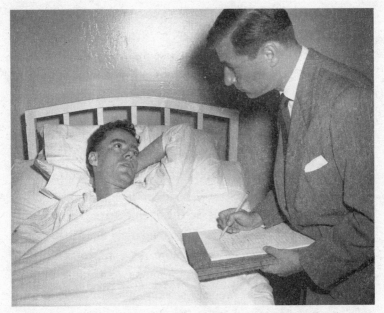

*Mitchell Cohen questions Howard Unruh in a hospital bed,
September 7, 1949.*

of indictable crimes, from gambling rackets, to robberies, to
teens illegally buying beer.

The office was the opposite of bustling. None of the detectives were around, and the quiet cast a strange pall over the
place. Then the phone rang. Larry Doran, chief of detectives, was
on the line. He told Cohen that a local man had gone "berserk
on River Road and was shooting people," which was why every
police officer was out of the office. He also told Cohen that Unruh was alive and in custody after his twenty-minute rampage.

Cohen walked over to the police station to interview the mass
shooter and found him cooperative. "It was a horrible, revolting
narrative," Cohen recalled in a 1974 interview. "He really gave
it cold, cut-and-dry. There was no attempt to conceal or be furtive. He didn't seem to experience the normal relief of getting

it off his chest. There was no remorse, no tears. There was a lack of all emotion."

Over the two or so hours he spoke with Cohen, Unruh was concealing something. When Cohen realized what it was, he was stunned. "What really convinced me that [Unruh] was terribly insane was when he got up after two hours and his chair was covered with blood. . . . He had been shot and wasn't even aware of it." Unruh was sent to a nearby hospital to recuperate, and Cohen interrogated him further upon his recovery from the bullet wound.

One month after the massacre, Cohen released the psychiatric reports he'd ordered on Unruh to the public. Unruh had been ruled clinically insane, and therefore not competent to stand trial. And so the deaths of thirteen people and the injuries of many more were never properly accounted for in court. Unruh didn't go free. He would spend the rest of his life in mental institutions in and around Trenton. But for those who survived the massacre, who attended hearing after hearing to ensure Unruh was never released, it did not seem like proper justice. He died in 2009 at the age of eighty-eight, just one month after Charles Cohen, the last survivor of the massacre, died.

Unruh's "Walk of Death" also seemed to foreshadow Camden's deeper decline. "It's something you never really forget. . . . You take extra precautions to protect your family and your property," Paul Schopp, a former director of the Camden County Historical Society, said in an interview to mark the sixtieth anniversary of the mass shooting. "He didn't just rob them of their lives. He robbed them of their essence." The trauma of a mass shooting, and a collective desire to forget, seems like the true beginning of Camden's downward slope.

TWELVE

Across America by Oldsmobile

Vladimir Nabokov finished the 1948–1949 academic year at Cornell University in a state of irritation. He hadn't found much time to write. He fumed over cuts and changes made without his permission by the *New York Times Book Review* to his review of Jean-Paul Sartre's *La Nausée*, which he had submitted in March. His finances were depleted: Nabokov hadn't budgeted for unexpected housing costs and the added expense of Social Security (what he termed "old-age insurance") taken out of his monthly salary. And he was exhausted from teaching a full load of English and Russian literature undergraduate classes, exacerbated by the extra work he'd inflicted upon himself by translating a pivotal Russian poetic masterpiece, "The Song of Igor's Campaign," for one of those classes.

Nabokov had, at least, completed another two chapters

of his memoir, *Conclusive Evidence,* both of which were published later that year in the *New Yorker.* He did love teaching,
and Cornell proved to be more amenable to his idiosyncrasies
than Wellesley. But he couldn't resist complaining: "I have always more to do than I can fit into the most elastic time, even
with the most careful packing," he wrote his friend Mstislav
Dobuzhinsky in the spring of 1949. "At the moment I am
surrounded by the scaffoldings of several large structures on
which I have to work by fits and starts and very slowly."

Lolita, which he still thought of as *The Kingdom by the Sea,*
was less a work in progress than a seed in Nabokov's mind, one
that wasn't quite ready to germinate. Perhaps he would make
a beautiful work on his summer trip—another cross-country
jaunt with Véra and Dmitri. They said goodbye to the Plym-
outh that had carried them all the way to Palo Alto, California,
in 1941, and hello to a used black 1946 Oldsmobile. Dorothy
Leuthold, who had shared the driving with Véra eight years
earlier, wasn't available, and neither were two other friends,
Andree Bruel and Vladimir Zenzinov. But one of Nabokov's
Russian literature students, Richard Buxbaum, volunteered,
and the Nabokovs picked him up at Canandaigua on June 22.

Their first destination was Salt Lake City, where Nabokov
was to take part in a ten-day writers' conference at the University of Utah starting on July 5. But their westward journey almost ended a few miles from Canandaigua, when Véra
changed lanes on the highway and narrowly missed plowing
into an oncoming truck. Pulling over, she turned to Buxbaum
and said: "Perhaps you'd better drive."

With Buxbaum now behind the wheel, the group traveled
south of the Great Lakes and across Iowa and Nebraska. The
Nabokovs spoke Russian and encouraged Buxbaum to do the
same, chiding him when he lapsed into English. Vladimir was

never without his notebook, ready to record all observations, however minuscule, of quotidian American life on the road, be it overheard conversation at a restaurant or vivid impressions of the landscape. They arrived in Salt Lake City on July 3, two days before the conference's start, and were lodged at a sorority house, Alpha Delta Phi, where the Nabokovs had a room with a private bath—a pivotal part of his participation agreement.

The conference introduced Nabokov to writers he might not have otherwise met, including John Crowe Ransom, the poet and critic who founded and edited the *Kenyon Review;* and Ted Geisel, a few years away from children's book superstardom as Dr. Seuss, whom Nabokov recalled as "a charming man, one of the most gifted people on this list." He also got reacquainted with Wallace Stegner, whom he'd first met at Stanford. Nabokov and Stegner spent the conference debating each other in the novel workshops and in the off-hours playing doubles on the tennis courts, with their sons as partners.

Nabokov did not have much time to idle, though. He taught three workshops on the novel, one on the short story, and another on biography. He took part in a reading with several poets, and repurposed an old lecture on Russian literature under a new title, "The Government, the Critic, and the Reader." When the conference ended on July 16, he, Véra, Dmitri, and Richard Buxbaum headed north to the Grand Tetons in Wyoming.

Nabokov, once more, was game to hunt more butterflies. But Véra was worried. The Teton Range, she had heard, was a haven for grizzly bears. How would Vladimir protect himself against them carrying a mere butterfly net? Nabokov wrote to the lepidopterist Alexander Klots for advice; Klots assured him that Grand Teton was "just another damned touristed-out National Park." Any danger would come from clueless visitors, not ravenous bears.

Nabokov re-creating the process of writing Lolita *on note cards.*

From there the quartet headed to Jackson Hole, where Nabokov wanted to look for a particularly elusive subspecies of butterfly, *Lycaeides argyrognomon longinus*. On the way the Oldsmobile blew a tire. As Dmitri and Richard started changing it, Nabokov said, "I'm no use to you," and spent the next hour catching butterflies. They arrived the following day, around July 19. For the next month and a half, the Nabokovs' home base was the Teton Pass Ranch, at the foot of the mountain range. Nabokov's hunt for his coveted butterfly subspecies proved successful.

The six-week stay was not without rough moments, though. Dmitri and Richard Buxbaum decided to try climbing Disappointment Peak, next to the Grand Tetons' East Ridge. The climb to the top, seven thousand feet above base level, was straightforward at first. Then Dmitri, with the overconfidence befitting a fifteen-year-old boy, decided they should switch to a more difficult path, one that required extra equipment they lacked. Realizing they would get stuck up there if they carried on, they turned around, but hours passed—and the sun nearly set—by the time they made it back to Vladimir and Véra, who were understandably frantic.

Buxbaum hitchhiked home at the end of August. The Nabokovs, with Véra now driving, ventured northeast to Minnesota, then up to northern Ontario, for more butterfly collecting, before they finally arrived in Ithaca on September 4. Nabokov had three courses to teach that fall, but had attracted only twenty-one students, combined, a workload greeted with suspicion by his fellow professors. Even so, Nabokov wanted more money.

Cornell's head of the Literature Department, David Daiches, received Nabokov's request and offered a deal: he would approve a salary raise if Nabokov took over teaching the European fiction course, Literature 311-12. Nabokov could shape the curriculum as he saw fit and pick the authors he liked most. Nabokov said yes. He began right away, scribbling notes on the back of Daiches's letter for the course that would define his Cornell career for the next decade.

And in spare moments, Nabokov began at last to shape the novel that had lived in his head for so long.

NABOKOV'S RAMSDALE IS not Camden. The made-up town where Humbert Humbert insinuates himself into the lives of Charlotte and Dolores Haze is most likely located in New England, which is why Dolores is deposited in a school in the Berkshires, and why both older man and younger girl seem to know the area well. Nabokov gleaned this knowledge during his years living in Cambridge, Massachusetts. But the names of the towns are similar, as is the Linden Street of Sally Horner's childhood home and the Lawn Street where the Haze women live. Both towns shared a white, middle-class bucolic atmosphere. As would happen again and again, the Sally Horner story parallels *Lolita* in all sorts of surprising ways.

Humbert Humbert came to Ramsdale by design, but moved into the Haze home at 342 Lawn Street by accident. He meant to stay nearby with the McCoos, parents to "two little daughters, one a baby the other a girl of twelve." Lodging there, he assumed, would allow him to "coach in French and fondle in Humbertish." But when Humbert gets there, he finds out that the McCoos' house has burned down, and he must find someplace else to live.

He is not pleased to be shuffled off to a "white-frame horror . . . looking dingy and old, more gray than white—the kind of place you know will have a rubber tube affixable to the tub faucet in lieu of a shower." Humbert is further irked upon overhearing Charlotte Haze's contralto voice ask a friend if "Monsieur Humbert" has arrived.

Then she comes down the steps—"sandals, maroon slacks, yellow silk blouse, squarish face, in that order"—tapping her cigarette with her index finger. In Humbert's estimation, Charlotte isn't much: "the poor lady was in her middle thirties, she had a shiny forehead, plucked eyebrows and quite simple but not unattractive features of a type that may be defined as a weak solution of Marlene Dietrich." Then he spies Dolores, and it is as if he saw his "Riviera love peering at me over dark glasses," two and a half decades after his prepubescent romance with Annabel, as if the years "tapered to a palpitating point, and vanished." Now that the true object of Humbert's obsession has revealed herself, Charlotte becomes a nuisance to be manipulated and endured.

When Charlotte sends Humbert a letter, confessing she is "a passionate and lonely woman and you are the love of my life," he senses opportunity: marry Charlotte to gain access to Dolores. Another line in Charlotte's letter stands out: that if Humbert were to take advantage of her, "then you would be a criminal—worse than a kidnapper who rapes a child." Charlotte, bafflingly, concludes that if he showed no sign of romantic interest in her, and remained in her home, then she would take it that he was "ready to link up your life with mine forever and ever and be a father to my little girl." (Since we're always in Humbert's head, we only have his word that Charlotte wrote this.)

The American widow and the European widower marry in haste while Dolores is away at summer camp. It is, suffice to say, a bad match. What galls Humbert the most is what he describes as Charlotte's vituperative attitude toward her daughter. The words Charlotte underlined in her copy of *A Guide to Your Child's Development* to mark her daughter's twelfth birthday: "aggressive, boisterous, critical, distrustful, impatient, irritable, inquisitive, listless, negativistic (underlined twice) and obstinate."

Humbert has already decided to murder Charlotte and stage it as an accident, perhaps at a thinly populated beach they visited ("The setting was really perfect for a brisk bubbling murder"). His simmering rage boils over when Charlotte informs him that she intends to send Dolores to boarding school at Beardsley, so that the two of them can take a trip to England. He resists. They argue. Then she reveals to him that she has read his notes, and knows the truth about him: "You're a monster. You're a detestable, abominable, criminal fraud. If you come near—I'll scream out the window. Get back!"

Humbert exits the house. He notes that Charlotte's face is "disfigured by her emotion" and remains calm. He goes back into the house. He opens a bottle of Scotch. Then, quietly, he begins to gaslight her. "You are ruining my life and yours. Let us be civilized people. It is all your hallucination. You are crazy, Charlotte. The notes you found were fragments of a novel."

Charlotte runs back to her room, claiming she has a letter to write. Humbert makes her a drink—or so he says—while she is gone. He realizes she is not, in fact, in her room. The telephone rings. "Mrs. Humbert, sir, has been run over and you'd better come quick." Fate has played a trick. Instead of becoming the potential savior of her daughter's virtue, Charlotte ends up dead. It's also played a fast one upon Humbert: he was all set to become a murderer.

THE DESCRIPTION OF CHARLOTTE HAZE as Dietrich-lite sounds a jarringly familiar bell when I look at pictures of Ella Horner from when her daughter disappeared in 1948. She was forty-one and often wore her hair pulled back, sometimes in a bun (a "bronze-brown bun"), and plucked her eyebrows over eyes that tended to disappear into their creases. Her other facial features—strong jawline, prominent nose, pronounced cheekbones—were reminiscent of Marlene Dietrich before she emigrated from Germany and became a Hollywood star. Based on the photographs of her that I've seen from before and after Sally's kidnapping, I imagine when Ella smiled, it didn't often reach her eyes. (Humbert on Charlotte: "Her smile was but a quizzical jerk of one eyebrow.")

There is another similarity, coincidence or otherwise, that

ties Charlotte Haze to Ella Horner: the device of marriage to gain access to their daughter. For the fictional Humbert Humbert it was a real gambit. For Frank La Salle, it was a delusion he created to explain why he took Sally away from her mother. It was a ruse Sally had to live by in order to survive being with him in Atlantic City and Baltimore, and she would have to endure it for quite a while longer.

THIRTEEN

Dallas

Frank La Salle took Sally Horner aside one day in March 1949 and broke the news that they were leaving Baltimore. He told her that the FBI had assigned him a new case, one that required him to move southwest to investigate. By then she'd been with him for nine months. Sally did not know, and could not know, that the real reason they were leaving Baltimore was that Camden County prosecutor Mitchell Cohen had indicted La Salle on the more serious charge of kidnapping on March 17. The new indictment meant La Salle could face between thirty and thirty-five years for taking Sally. Police had not located the pair, but this charge, on top of the original indictment, promised greater scrutiny, more resources for the hunt, and a better probability of arrest. Baltimore was no longer safe, nor was the entire East Coast. Instead of flushing La Salle out, the new charge caused him to run.

The journey from Baltimore to Dallas is approximately 1,366 miles. Today, by car, it would take about twenty hours to drive, on I-81 and I-40. Neither of those highways existed in 1949. La Salle and Sally likely drove south on U.S. 11, traveling all the way to the highway's end in New Orleans, Louisiana, before switching over to U.S. 80, arriving in Dallas less than two hundred miles later. However they traveled, Sally and La Salle got to Dallas around April 22, 1949. For the next eleven months, she and Frank continued to play father and daughter, sticking to the cover story that he had taken Sally away from her wayward mother to provide her with a more stable upbringing. None of their new neighbors seemed to question this. At least, not right away.

They moved into a quiet, well-kept trailer park on West Commerce Street, about four hundred feet from Dallas's bustling downtown core. The park was designed like a horseshoe, with trailers—including one that La Salle bought on the premises—dotted all along the curve. The park could hold as many as a hundred motor homes. The mothers mostly stayed home and the fathers worked as farmhands, for steel companies, or at gas stations. Neighbors were closer in the trailer park than they had been in Baltimore. They could pay more attention to the pair, and get to know Sally—or think they did.

La Salle had changed their names again. Sally was no longer known as Madeline LaPlante, but as Florence Planette. Oddly, it isn't clear whether La Salle also used the "Planette" alias. One of their new neighbors, Dale Kagamaster, who ended up working with La Salle, knew him as LaPlante. Frank also told people that he was widowed, a change from the divorced father cover story he used in Atlantic City and Baltimore.

The trailer park was owned by Nelrose and Charles Pfeil, who'd bought it a year earlier, after moving to Dallas from Akron, Ohio, with their three sons. Tom, the eldest, was nine years old when Frank and Sally arrived at the trailer park. He did not recall the name "LaPlante," but thought "La Salle" seemed familiar. He also remembered "Florence's" father as aloof, cold, standoffish. "I understand why, now," Pfeil told me. "He had to be suspicious of everyone around him." Tom had dim memories of Sally. "I don't know if I could tell you I remember much of her except for talking a time or two. I was nine. I just wanted to play ball."

As in Baltimore, La Salle got a job as a mechanic, but kept Sally in the dark about what he was really doing all day. He also enrolled Sally in another Catholic school. This time it was Our Lady of Good Counsel Academy at 210 Marsalis Avenue in the neighborhood of Oak Cliff, about a seven-minute drive away from West Commerce Street. Like Saint Ann's in Baltimore, Our Lady of Good Counsel no longer exists, having been absorbed into Bishop Dunne Catholic School in 1961. None of its records have survived. And also like Saint Ann's, the school was in a predominantly white, middle-class neighborhood that is no longer so, thanks to suburban migration, systemic inequality, and poverty.

Sally likely kept to a routine similar to the one she had in Baltimore. A bus ride to Our Lady of Good Counsel, where she began the day with morning prayers. Schoolwork wasn't a breeze, but her grades were generally good: a copy of Sally's report card from between September 1949 and February 1950, which La Salle kept after she brought it home, showed she received primarily As and A-minuses, with the occasional B, in geography and writing. The only time she received a lower grade—a C-plus, in languages—was in her final month at the school.

At first their Commerce Street neighbors didn't see anything amiss. Sally appeared to be a typical twelve-year-old living with her widowed father, albeit one he never let out of his sight except to go to school. Sally never displayed despair or asked for help. La Salle wouldn't let her.

Her neighbors thought Sally seemed to enjoy taking care of her home. She would bake every once in a while. She had a dog, one she apparently spoiled. La Salle provided her with a generous allowance for clothes and sweets. She would go shopping, swimming, and to her neighbors' trailers for dinner—sometimes with La Salle, and other times by herself, when he told her he was working the case for the FBI.

Dale Kagamaster's wife, Josephine, thought Sally was a well-adjusted girl. "There were several times we noticed the need for the love and care of a mother but we both felt that the father was doing a good job of providing better living conditions for [her]." The consensus about Sally and her "father" was that they "seemed happy and entirely devoted to each other." Maude Smillie, who was living in a nearby trailer, seemed bewildered by the idea that Sally had been a virtual prisoner: "[Sally] spent one day at the beauty parlor with me. I gave her a permanent and she never mentioned a thing. She should have known she could have confided in me."

Nelrose Pfeil was quoted in a court document several years later saying something similar: "Sally was in my home many times a day and she had access to several phones should she choose to use one. Sally had plenty of time to talk to me about being kidnaped [sic] if she had wanted to and I am sure she knew me well enough to know if she had said anything like that I would have helped her." The only time La Salle kept Sally from playing with other children, according to Pfeil's statement, "was when the person's character was in question."

It appears the Pfeils, the Kagamasters, and other neighbors

bought La Salle's cover story about Sally. They did not notice anything amiss, even for the ten-day period when Sally dropped off the radar and didn't attend school. She'd suffered an appendicitis attack, one that required her to undergo an operation and spend three nights at the Texas Crippled Children's Hospital (now the Texas Scottish Rite Hospital for Children). The other seven days, presumably, she spent at home, recuperating.

Something did change in Sally, though, after the operation. She grew more pensive. Josephine Kagamaster observed that the girl did not move like a "healthy, light-hearted youngster." She'd heard La Salle say the girl "walks like an old woman."

ON THE SURFACE, Sally acted as free as she had been in Camden, before Frank La Salle took her away from everyone she loved. She might have been left alone for long stretches at a time, stayed late at a neighbor's house watching television, and been on her own in the hospital for several nights. But if she told the truth, who would believe her story? Who would believe she had been abducted when, to all appearances, Frank La Salle was her father, and a loving one at that? And even if someone *did* believe her, could they help, or would they put Sally in greater peril?

Later, Josephine Kagamaster, Nelrose Pfeil, Maude Smillie, and others said they would have helped Sally had she chosen to confide in them. But they made such declarations with the benefit of hindsight, months or years after Frank La Salle's diabolical crimes were exposed to the public. At the time, they were living ordinary and happy lives. The idea that a young girl and an older man would be in a cruel parody of a father-daughter relationship seemed inconceivable, unimaginable. And no matter what they believed about what they would have done, Sally did not confide in these neighbors. She did not feel she could trust them.

But Sally did talk to someone, a woman named Ruth Janisch, and she believed what the girl had to say. Though Ruth's motivations were more complex than anyone knew, her belief in the girl eventually emboldened Sally to make the most important decision of her life.

FOURTEEN

The Neighbor

Ruth Janisch and her family arrived at the Commerce Street trailer park around December 1948. They had spent most of the 1940s traveling a particular geographic loop, following the employment her husband, George, found repairing televisions or working in bowling alleys. It began in San Jose, where Ruth and George met and married, then moved up to Washington, where she'd grown up, tracked east to Minnesota, the home of George's parents, and finally on to Texas, situated more or less in between. The Janisches bought a caravan somewhere along the way and made it the family home.

Periodically, the trailer ran into trouble. On Thanksgiving 1948, it broke down on the way to Dallas, somewhere in the desert. New Mexico, perhaps, or Arizona. George and his elder stepson, Pat, went looking for help, leaving the rest of the family stranded by the road. Ruth and her other children—another boy from an earlier marriage, and two girls sired by George—figured that if they were stuck by the road, they might as well have Thanksgiving dinner while they waited.

They fetched chairs from a closet in the trailer and set up outside. Ruth cooked up an impromptu meal of pancakes and beans, which she served inside the broken-down trailer. The children lined up to get their meal and then ate outside in the baking desert sun. Ruth warned the children not to stay outside for too long. She was nervous that rattlesnakes might bite them if the kids lingered.

Eventually George and Pat returned with the part they needed to fix the trailer, and they drove on to Dallas, setting up camp at the Commerce Street site. A few months later, in April 1949, a man in his fifties and a girl he said was his daughter moved into the trailer next door. The Janisch girls immediately took to the girl, who introduced herself as Florence Planette. She was twelve, practically grown up, but she was willing to give them her attention. The little girls were five, six, and seven, and regarded her with a mixture of awe and envy.

Ruth may well have regarded the girl's father with extra-marital interest. That's her children's theory now. Whatever her motives, Ruth noticed something askew in the relationship between Sally Horner and Frank La Salle that had eluded everyone else who interacted with them. What Ruth saw between the older man and the young girl spurred her to the single gesture that defined her as a decent human being, an act she would relive for the rest of her days and memorialize in scrapbooks. That act did not make her a heroine in the eyes of her children. But it would bring her a level of attention she spent the rest of her life trying to find again.

Ruth Janisch may have been suspicious of Frank La Salle

because she wasn't in the habit of trusting people. She craved love she never found. She got pregnant so often she was in a perpetual state of exhaustion, dealing with babies and children. George always found work, but the money he brought in was hardly enough for an ever-expanding family. When her children misbehaved, it was all too easy for Ruth to fall back into the patterns she learned as a child, berating them the way her mother had berated her, telling them they were worthless, useless, or worse.

Ruth Janisch, ca. 1940s.

Her bitter outlook took hold upon leaving Washington State to marry her second husband, Everett Findley. (Ruth later said her first marriage, at sixteen to a man whose name she failed to remember, didn't count.) The former Ruth Douglass was eager to flee her mother, Myrtle, whose cuts were always unkind, and her father, Frank, whom her children later grew fond of but whom Ruth, in her cups, recalled as being "not so innocent." The children were never sure if Ruth was referring to her father's penchant for drink or something uglier.

After their marriage, Ruth had followed Findley, a man more than twice her age, to San Jose, and bore him two sons. She met husband number three, George Janisch, sometime after the dissolution of her marriage to Findley. George hailed from Minneapolis; he was short and slight, and his blond hair and fair appearance befitted his Scandinavian heritage.

He'd moved west for work and to escape the harsh Minnesota winter.

George and Ruth ran off to Carson City, Nevada, to wed on October 24, 1940. Perhaps they married for love. Not long before he died, George confided in one of his daughters that before their wedding, Ruth was a "good girl." But afterward, according to George, she changed, and he admitted that it was his fault.

It wasn't enough for George to sleep with his wife. He had to sleep with other men's wives, too. Ruth herself had taken up with him while he was married. Since George was fine if Ruth slept with the leftover husbands, she wasn't about to say no. The fact was that Ruth had a craving for men that would persist for the rest of her life.

The extramarital doings damaged the already tenuous bond between the Janisches, which had been frayed by having three daughters in quick succession. The couple seemed to bring out the worst in each other. One particularly clever, or insidious, way Ruth and George tested each other was with the naming of every new child. Each baby received a first name either spouse liked. The middle names, however, were those of former lovers. Nine children later, Ruth and George split up. He would marry twice more; she married ten times in total, with lovers scattered in between.

By 1949, Ruth was thirty-three (though would only admit to thirty-one) with a husband she couldn't help needling and at the mercy of that perpetual pregnancy-birth cycle. She still had most of her looks, with dark hair curling about her face, full, pointed breasts, a strong nose and wide-lipped mouth. Every new child added another dose of bitterness at her lot in life, and the family's poverty.

But there was something about Sally Horner that Ruth

could see clearly. The way the girl shuffled after coming home from an extended hospital stay after an appendectomy. The way Sally's smile didn't reach her eyes. The closeness between Sally and Frank that did not strike the right note. "He never let Sally out of his sight, except when she was at school," Ruth later recounted. "She never had any friends her own age. She never went any place, just stayed with La Salle in the trailer." She thought La Salle seemed "abnormally possessive" of the girl he said was his daughter. Ruth tried to cajole Sally, still recovering from her appendectomy, to tell her the "true story" of her relationship with La Salle. Sally wouldn't open up.

In early 1950, the Janisches packed up their trailer and drove west. Work had dried up for George in Dallas, and he figured he might have better luck in San Jose, which had proven lucky in the past. Once the family, larger by two more children, landed at the El Cortez Motor Inn—perhaps at their exact prior parking space—Ruth wrote to Frank saying that he and Sally should follow them to California. There's work to be had here, she said. He and Sally could be their neighbors again.

La Salle agreed. Perhaps he had some other pressing reason to abandon Dallas. Maybe he sensed that Sally was distancing herself from him and another move might keep her closer. Whatever the reason, La Salle pulled Sally out of school in February 1950 and they drove the house trailer attached to his car from Dallas to San Jose. Just as in Baltimore, and Atlantic City before it, La Salle had decided he and Sally needed to be on the move. And just as before, Sally had no say in the decision. She did what Frank La Salle told her to do. But his mood was different on the day they headed west. This time they were running toward opportunity, not running from the law.

Sally and La Salle's journey to San Jose took at least a week, if not more. He drove the trailer through Texas, going around the border of Oklahoma, then through New Mexico, Arizona, and Southern California, before moving up the South Bay to their final destination, the farthest Sally had ever been from Camden. She would never venture this far again. Sally had been La Salle's captive for nearly two years, since she was just eleven. She felt his presence at every turn, even when she was alone and seemingly free to do what she pleased. How trapped she must have felt to be in such close quarters to him as they spent that week or ten days on the road.

If Sally had allowed herself to let her mind roam, she might have given in to feelings of despair, or to anger over what La Salle had taken away from her. Or perhaps she was focused on how vital it was for her to survive. After days in the car and nights in the trailer parked at a rest stop, eating at diners, one after another, the emotional toll on her must have been considerable.

On the West Coast, Northern California in particular, palm trees lined broad boulevards where cars had room to move instead of getting jammed up like they did back home. Police in uniform shorts patrolled the streets on motorcycles. The air was far less humid than in Dallas, or even on the East Coast. But the prospects of betterment that had enticed La Salle, and so many others before him, were not on Sally's mind. She had a great many other things to think about.

By the time Frank La Salle pulled the house trailer into the El Cortez Motor Inn on Saturday, March 18, Sally Horner felt able to reckon with the changes roiling inside her. She'd already made a significant first step. Before leaving Dallas, she'd mustered up the courage to tell a friend at school that her relationship with her "father" involved sexual intercourse. The friend told Sally her behavior was "wrong" and that "she ought to stop," as Sally later explained. As her friend's admonishment sank in, Sally began refusing La Salle's advances, but kept up the illusion he was her dad.

For so long she felt she had to stay silent, or to accept what the man posing as her father said was the natural thing to do between them. All this time she opted to give in because it seemed the surest path to survival. Now Sally felt freer in a small way. Not free—she was still in La Salle's clutches, and could not see a way to escape. But she could say no now, and he didn't punish her like he had in the old days. Perhaps he looked at Sally, a month shy of her thirteenth birthday, and saw a girl aging out of his tastes. Or perhaps he trusted that Sally belonged to him so completely he no longer needed to use rape as a means of physical and psychological control.

What she knew now was that her relationship with Frank La Salle was the opposite of natural. It was against nature. It was wrong.

FRANK LA SALLE needed to find work. Several days after landing at the trailer park, La Salle abandoned his car—perhaps it needed repairs after so many days on bumpy, unevenly paved highway roads—and took the bus two miles into town to look for a job. Sally was already enrolled in school, and may have attended as many as four days of classes. She did not attend class that morning, though. By staying away, Sally changed the course her life had traveled on for the past twenty-one months.

FIFTEEN

San Jose

On the morning of March 21, 1950, Ruth Janisch invited Sally Horner over to her trailer. She knew Frank La Salle wouldn't return from his job search for several more hours, and sensed the girl might open up to her. All it would take was the right push at the right time. If Ruth didn't seize the opportunity now, she never would. Gently, she coaxed more honesty out of the young girl. Before, in Dallas, Sally wouldn't budge. This time, in San Jose, she did.

Sally confirmed Ruth's suspicion that Frank La Salle was not, in fact, her father, and that he'd forced her to stay with him for nearly two years. She said she missed her mother, Ella, and her older sister, Susan. Sally told Ruth she wanted to go home.

Ruth absorbed what the girl told her. Though she had been suspicious of the relationship, she never imagined that La Salle had kidnapped Sally. Then she sprang into action. She beckoned Sally over to the telephone and showed her how to make a long-distance phone call. Sally had never done so before.

Sally dialed her mother's number first, but the line was disconnected; Ella had lost her seamstress job in January and, while unemployed, could not afford to pay the bill. Next, she tried her sister, Susan, in Florence. No one answered the house phone, so Sally tried the greenhouse next.

Her brother-in-law, Al Panaro, picked up.

"Will you accept a collect call from Sally Horner in San Jose, California?" the operator asked.

"You bet I will," Panaro replied.

"Hello, Al, this is Sally. May I speak to Susan?"

He could barely contain his excitement. "Where are you at? Give me your exact location."

"I'm with a lady friend in California. Send the FBI after me, please! Tell Mother I'm okay, and don't worry. I want to come home. I've been afraid to call before."

The connection was poor, and Al had a hard time hearing his sister-in-law. But he heard enough to get the trailer park address down on paper, and to assure Sally he would call the FBI. She just had to stay exactly where she was.

Then Panaro passed the phone over to Susan, who was with him in the greenhouse. She was flabbergasted that her younger sister was alive, and on the telephone line. She also urged Sally to stay put and wait for the police.

After Sally hung up, she turned to Ruth, her face drained of color. She looked ready to collapse. She kept saying, over and over, "What will Frank do when he finds out what I have done?"

Ruth spent the next little while keeping Sally calm, hoping the FBI, or even the local police, would show up soon and arrest Frank. Sally, anxious, thought she should go back to her own trailer to wait for the police. Ruth let her go, hoping it would not be for too long.

AFTER SPEAKING with his sister-in-law for the first time in nearly two years, Al Panaro immediately called the Camden County Police Department. He asked for Detective Marshall Thompson, the man who'd been investigating Sally's disappearance exclusively for more than a year. But Thompson worked the night shift and was home in bed when Panaro's call came in. William Marter, another detective, answered.

Marter was the one who relayed Sally's whereabouts to the New York FBI office. He warned them to proceed with caution around La Salle. He had eluded capture before, and they needed to be certain he would not escape again. Then the FBI rang the sheriff's office in Santa Clara County. Sheriff Howard Hornbuckle picked up, and soon learned that a girl abducted almost two years earlier was alive and well and in his jurisdiction.

Hornbuckle had been elected sheriff three years earlier. He was a local boy, a graduate of San Jose High School, and had attended the state college before he joined the police department in 1931. He had spent fourteen years on the force, as a detective and later a captain. He'd also moonlighted as a traffic safety instructor in his spare time, where he stressed the danger of cars and how too many young people died while at the wheel. A cautionary slogan he coined—"Death Begins at 40"—even got picked up by the wire services and circulated nationally for a while.

Santa Clara County had its fair share of crime. Hornbuckle's own predecessor was indicted on gambling and bribery charges, and more recently the brutal murder of a high school girl had garnered headlines. But this situation was extraordinary. While many in local law enforcement got their hackles up when the FBI called, Hornbuckle did not. The case of a young girl so far from home was no time to get your nose out of joint. The FBI and the sheriff's office would work together on this.

When Hornbuckle sent his deputies to the trailer park on Monterey Road, federal agents were already on their way. The fleet of cops, local and national, sped to the El Cortez Motor Inn. Three men from the sheriff's office, Lieutenant John Gibbons and Officers Frank Leva and Douglas Logan, found Sally, alone, in La Salle's trailer.

"Please get me away from here before he gets back from town," she said, terror winning out over relief for the moment. What if he returned before she could get away from the trailer park? What if he tried to take her again? And if he did, what if he did things to her she didn't want to think about?

But this time she was in the hands of the real police and the real FBI, not the pretend agent, Frank La Salle. These cops promised Sally she was safe. La Salle would not be able to take her or touch her again. Three deputies whisked her to a detention center in the city, run by Matron Lillian Nelson. Once she was settled there, the remaining local and federal police waited for La Salle to return.

Lieutenant Gibbons at first held back from questioning Sally. "She's too shaken up," he told reporters a few hours later when they pressed him for details. But when Sally calmed down, Sheriff Hornbuckle led her into an interview room where she told him what had happened, and where she had been all this time. Hornbuckle listened, with patience, as Sally told him the whole terrible story. At first she gasped, sobbed, and cried. The hysterics were understandable, and the sheriff did not hurry her.

Then, at last, Sally found her voice. She started at the beginning, describing how La Salle caught her trying to steal a notebook on a dare at the five-and-dime. How he said he was an FBI agent and that she was "under arrest." How scared she was, and then how relieved when he let her go. How he found her again several months later, coming home from school. And how he told her she could avoid reform school only if she went away to Atlantic City with him, telling her mother he was the father to her friends, "because the government insisted I go there."

Sally confirmed that she and La Salle had lived in Baltimore for eight months before moving on to Dallas, and had only just arrived in San Jose. The entire time he held on to her, La Salle told Sally "that if I went back home, or they sent for me, or I ran away, I'd go to prison. The government ordered him to keep me and take care of me, that's what he said."

Hornbuckle then had to ask Sally the toughest question: whether La Salle had forced her to have sex with him during their nearly two years on the road. He phrased it delicately, asking if Sally had "been intimate" with La Salle. She denied it. But later, after a doctor's examination, she confessed the truth. "The first time was in Baltimore right after we got there. And ever since, too." And then in Dallas, she said a "school chum of mine" told her that what she was doing with Frank was "wrong, and I ought to stop. I did stop, too."

She said La Salle was "mean and scolded me a little, but the rest of the time he treated me like a father." Sally also said he had carried a gun for a time, in keeping with his pose as an FBI agent, but she thought La Salle had left it behind in Baltimore.

Sally was emphatic that La Salle was not her father. "My real daddy died when I was six and I remember what he looks like. I never saw [La Salle] before that day in the dime store."

Once she began to talk, she could not stop. Until finally, pausing for breath, she said, "I want to go home as soon as I can."

IT'S NOT CLEAR if Frank La Salle found gainful employment that morning in San Jose. When he stepped off the bus and walked back to the trailer just after one o'clock in the afternoon, dozens of police officers surrounded him before he could reach his front door. They'd been hiding behind other trailers. Deputies from the sheriff's office. FBI agents. Local San Jose cops. All present because of a chain of events that began as soon as Sally Horner hung up the phone. La Salle did not fight, but instead surrendered quietly.

At the San Jose jail, La Salle grew more animated. He denied abducting Sally. He insisted he was her father and that her mother "has known where I am and where the girl is every day since I've been gone." He claimed that his wife was dead, and when confronted with the truth of his divorce from Dorothy Dare, denied his earlier claim. He also denied possessing newspaper and magazine photographs of young girls, defaced with obscene messages, as well as crude drawings laced with profanity, which FBI agents discovered in the trailer upon his arrest. La Salle elaborated on his alternate reality about Sally. "I took her when she was a little thing. . . . I am the father of six kids, three by this wife (Mrs. Horner) and three by another wife. I didn't take [Sally] from Camden but from New York. It was four years ago, not two. She kept house for me and she had money and freedom." The authorities, La Salle claimed, could have found him "at any time." He had a business in Dallas, after all, and "always had cars registered in my name." When he was done protesting his innocence, La Salle refused to speak further.

"He's a tough, vicious character," said Lieutenant Gibbons.

ELLA HORNER WAS OVERJOYED and overcome by the news that her daughter was alive and had been found. So much so that at first, she could hardly speak. When she composed herself, she told the large crowd of reporters and photographers who had descended upon 944 Linden Street that she was chiefly concerned with Sally's safety. "I just want her back and to see her again. I am very thankful, and I will be a whole lot more thankful when I really see Sally."

She also repeated the sentiment she'd expressed to the press—and, perhaps, countless other times in private—back in December 1948, while Sally was still missing. "Whatever she has done, I can forgive her."

Later that day, a Camden *Courier-Post* reporter, Jacob Weiner, found Ella clutching a photo of Sally, the one that had been recovered from the Atlantic City boardinghouse in August 1948. "It seems so long ago, Sally, so long ago," Ella murmured, gazing at the picture of her daughter. In a stronger tone, but with her voice still shaking, Ella said: "I'm so relieved."

Ella repeated that Sally had been gone for nearly two years. "That's a long time," she said. "During that time, I didn't hear from her. No word. No postcard. No news of any kind."

About that June day when she allowed Sally to accompany Frank La Salle for a seashore vacation, she said, "I must have been very foolish . . . at least I know it now." She picked up the picture of Sally again. "Anyway, I let her go. I haven't seen her since. . . ."

Weiner asked Ella if she ever gave up hope that Sally

would be found alive. There were times, Ella said, where she felt "pretty hopeless" because "I always knew she had enough sense to call me or drop me a line." And yet Sally hadn't.

What did Ella think about Frank La Salle? "That man . . . ," she began, but her voice broke.

Susan was sitting with her mother during Weiner's interview, and picked up the thread. "I hope that man La Salle is properly punished. He should receive life imprisonment . . . or the electric chair."

Then Susan turned her thoughts to a second telephone conversation she'd just had with her younger sister. "I couldn't believe it was Sally I was talking to. It was wonderful." Her eyes filled with tears. "I can't wait to see her."

Sally had asked Susan how their mother was faring. She also asked after Susan's daughter, Diana, now nineteen months old.

"She looks just like you," Susan said, and Sally burst into tears.

TELEPHONES ARE A recurring motif in *Lolita*. The incessant ringing of the "machina telephonica and its sudden god" interrupts the narrative, as Humbert Humbert's psyche begins to fissure—the monster underneath waging war with the amiable surface personality he presents to the world. Telephones are also the means through which Humbert discovers Charlotte's accidental death, since he is too preoccupied with fixing her a drink to notice that she has left the house.

With Charlotte permanently out of the picture, he goes to pick up Dolores at Camp Q to break the news of her mother's death in his own special way—"all a-jitter lest delay might give her the opportunity of some idle telephone call to Ramsdale." After he picks her up, he takes Dolores to the Enchanted Hunters hotel, where he rapes her for the first time. The following morning, the telephone plays a pivotal role in binding the older man and girl together. Humbert had told Dolores that he was taking her to Charlotte, who he said was in the hospital in Lepingville. At a rest stop, Dolores asks: "Give me some dimes and nickels. I want to call mother in the hospital. What's the number?"

Sally on the telephone to her family in the hours after her rescue.

Humbert says, "You can't call that number."

"Why?" cries Dolores. "Why can't I call my mother if I want to?"

"Because," he says, "your mother is dead."

It is the news that totally breaks Lolita and puts her in Humbert Humbert's power. He knows it, too: "At the hotel we had separate rooms, but in the middle of the night she came sobbing into mine, and we made it up very gently. You see, she had absolutely nowhere else to go."

From there Humbert and Dolores begin their road trip, a journey that would take them thousands of miles across the United States. Deep into their trip, Humbert's paranoia grows as he suspects Dolores has confided the truth about him to Mona, a school friend suspicious of the relationship between the so-called father and daughter: "the stealthy thought . . . that perhaps after all Mona was right, and she, orphan Lo, could expose [Humbert] without getting penalized herself."

Dolores's first escape, after she yells "unprintable things" and accuses Humbert of murdering her mother and violating her, occurs as the phone rings and she breaks free of his grip on her wrist (in part echoing La Salle's grip upon Sally's arm at the Camden five-and-dime). That escape lasts only a few hours, and Humbert finds her "some ten paces away, through the glass of a telephone booth (membranous god still with us)."

After that, Dolores asserts her will as to where they should go next. And then, though the reader is not privy to it, she makes a final, mysterious call, presumably to Clare Quilty, to help her escape. Telephones, Humbert concludes, "happened to be, for reasons unfathomable, the points where my destiny was liable to catch." For Dolores, telephones are the means for her to find freedom from the abuser who has engulfed her life—just as a telephone call was for Sally Horner.

SIXTEEN

After the Rescue

Though Frank La Salle was in jail, it wasn't clear which law enforcement agency would have jurisdiction over him. There were the outstanding warrants for kidnapping and abduction from Camden County. But because La Salle had transported Sally across several states, it became a federal case. La Salle was charged with violating the Mann Act, for "allegedly taking the girl across state lines for immoral purposes."

On the morning of March 22, Camden County prosecutor Mitchell Cohen spoke with the San Jose police, including Sheriff Hornbuckle. After the thirty-minute call, he told reporters in Camden that he would convene a grand jury to indict La Salle on the outstanding warrants, and start extradition proceedings immediately.

La Salle seemed ready to fight his extradition to Camden,

but Cohen was undeterred. "Regardless of what La Salle says he will do about returning here, I am taking no chances," Cohen said. "I will start formal proceedings at once and get him back here as soon as possible." But the prosecutor had to wait on New Jersey governor Alfred Driscoll's approval, and there was a delay because Driscoll was out of town on a business trip.

That afternoon, in California, Commissioner Marshall Hall presided over La Salle's arraignment on the Mann Act charges. He set a $10,000 bond and scheduled a hearing for the following morning. La Salle retained Manny Gomez as his attorney, while Frank Hennessy was the federal prosecutor.

The hearing began at 10:30 A.M. on March 23. There Hennessy revealed that La Salle's birth name was Frank La Plante; if true, then at various points during Sally's captivity, she'd attended school using the first name of La Salle's biological daughter and his own real last name.

When police officers attempted to lead Sally into the courtroom, she resisted at first, frantic at the thought of seeing La Salle: "I'm afraid, I'm afraid," she cried.

May Smothers, a juvenile court matron, had accompanied the girl to court, and calmed her down. Sally finally entered the courtroom clutching Smothers's hand. She took a seat only four feet away from La Salle and stole furtive glances at him throughout the proceedings, looking away whenever she came close to breaking down. La Salle stared at her, impassive, saying nothing.

When Sally began her testimony, Commissioner Hall asked, "Are you afraid of anything? Is there anything you want?"

"I want to go home!"

"He can't hurt you," said Hall.

And so, once more, Sally described her ordeal, starting with the Camden five-and-dime and ending with the San Jose trailer park. She told the court how La Salle had forced her to have sex with him, the abuse only ending in Dallas. La Salle told his story again, too, continuing to insist that he was Sally's real father.

Commissioner Hall affirmed the $10,000 bond, and ordered La Salle transferred to the county jail in San Francisco.

The hearing also decided La Salle's jurisdictional fate: Hennessy told the court that the federal charges would eventually be dropped because the New Jersey state kidnapping charges took precedence. But for the time being, La Salle would sit tight. Even if he raised the full $10,000 bond, federal authorities "were confident they could hold [La Salle] on other charges until he could be extradited," reported the *Courier-Post*.

Sally returned to the San Jose detention center. At first, she was so anxious about La Salle possibly going free that she could hardly eat. Matron Smothers told the papers that Sally also "fretted a lot about whether her folks would want her after what happened." Sally was kept apart from the other detained juveniles because, an unnamed sheriff's official told the *Courier-Post*, "We have some pretty hardened kids here and we don't want Sally to come in contact with them."

Over the next few days, Sally grew more secure in the detention center. Matron Smothers took her shopping for new clothes, because in her estimation, Sally's old ones did not measure up: "The clothes she had at the [trailer park] were neat but shabby and very inadequate." Smothers said that Sally had also stopped worrying about whether her family would welcome her back. "All she's thinking about is getting home and what she'll do when she gets there."

The detention center felt "responsible for Sally's well-being until New Jersey's authorities arrive to take her home," said an unnamed sheriff's official. "We've had a number of offers from people in San Jose to take care of Sally until she's ready to go home, but we are positive no harm can come to her where she is now."

BACK IN CAMDEN, police continued to investigate another dangling thread: the mysterious "Miss Robinson" Sally said had accompanied her and Frank La Salle on the bus to Baltimore, after which she disappeared. Camden police tried to reconcile Sally's statement to Sheriff Hornbuckle with what they found in their own initial investigations. They had proof, after all, that Sally and La Salle had spent time in Atlantic City, in the form of unsent letters, photographs, clothing, and other material abandoned at 203 Pacific Avenue. Proof bolstered by the recollections of Robert and Jean Pfeffer, the young Philadelphia couple who had reported spending a summer day with Sally and La Salle.

No trace of the woman known as "Miss Robinson" was ever discovered by law enforcement. It remains another of the unresolved mysteries of Sally's captivity. I believe the woman existed, because I believe Sally. Just because police did not track the woman down, and that decades later I also could not find her, does not mean Sally made her up.

A CAMDEN GRAND JURY indicted La Salle for kidnapping and abduction at 2:20 P.M. on March 23, the same day as the hearing in San Francisco. Ella Horner testified in front of the grand jury. There's no record of what she said, but she was likely asked about why she put Sally on the bus to Atlantic City and whether La Salle was her daughter's biological father, as he claimed.

Mitchell Cohen sent a copy of the grand jury indictment

to the New Jersey governor to start the extradition process. A second copy of the proceedings, signed by Judge Rocco Palese, was airmailed to California to reinforce La Salle's detention. Cohen also received permission to bring both Sally and La Salle—separately—back to Camden, and to cover their travel expenses, as well as those of Camden city detective Marshall Thompson and county detective Wilfred Dube.

Cohen, Dube, and Thompson flew into San Francisco on Sunday, March 26. Over the next few days, Cohen received approval to extradite La Salle from Governor Driscoll in New Jersey as well as his California counterpart (and future chief justice of the Supreme Court) Earl Warren. Cohen also interviewed various residents of the trailer park. One was Ruth Janisch, who told Cohen she was willing to testify at La Salle's trial.

On Thursday, Sally was released from the San Jose detention center into Cohen's custody. Just after 8:40 A.M. Pacific time on Friday, March 31, Sally and Cohen boarded a United Airlines flight headed for Philadelphia. Sally wore a navy-blue suit, polka-dot blouse, black shoes, a red coat, and a straw Easter bonnet for her first-ever plane trip. She told Cohen how much she looked forward to seeing her family. She threw up only once, when the plane ran into turbulence just outside of Chicago.

Ella waited at the airport in the backseat of Assistant

Sally Horner and Mitchell Cohen board a Philadelphia-bound United Airlines flight, March 31, 1950.

Camden County Prosecutor (and future New Jersey governor) William Cahill's car. The rest of Sally's family, including Susan, Al, and their baby, Diana, arrived separately. Several other planes landed first, each one lifting Ella's spirits before crushing them again. "Why doesn't it come," Ella said, her face pressed against the car window. Sally's plane finally landed just after midnight, just over an hour late.

From the plane, Sally spotted her brother-in-law in the crowd. Sally wanted to get out right away, but Cohen told her to wait for the other passengers to leave first. Then she spotted her mother. "I want to see Mama!" she cried.

"All right, Sally," said Cohen. "Let's go."

Sally stood at the doorway for a moment, looking around. Then she spotted her mother running toward her, holding out her arms. Sally raced down the steps, her face lit up with joy and washed in tears.

Sally sees her mother, Ella Horner, for the first time in twenty-one months.

She and her mother clung to each other for several minutes, oblivious to the myriad flash-bulbs popping at them. At first, they were weeping too hard to speak. Then Sally said: "I want to go home. I just want to go home."

When they were safely in Assistant Prosecutor Cahill's car, Ella explained to Sally that she couldn't go home just yet. Instead, the authorities would take her to the Camden County Children's Shelter in nearby Pennsauken, New Jersey, where she had to stay "until the trial is over."

Sally leans on her mother's shoulder minutes after they are reunited.

After a short drive, their car arrived at the center, the Panaros following closely behind in a separate vehicle. Susan got out of the car at the same time as Sally.

"Susan!" Sally cried upon spotting her older sibling. Sally had been so overwhelmed by the sight of her mother, the photographers, and so many well-wishers that she hadn't realized her sister was part of the crowd.

"I kissed you at the airport but you didn't recognize me!" Susan said.

Then Sally realized her sister was holding a little girl in her arms. Sally reached for Diana, the niece she'd never met, and hugged her tightly. "Gee, she looks like pictures of me taken when I was a baby!"

Cohen, exhausted from the trip, gently informed the family that Sally needed to get some sleep.

In the days that followed, Ella was the only family member allowed to visit Sally at the Children's Shelter, to ensure the girl stayed in a calm frame of mind before and during the trial. Fortunately, Sally got along well with the matron. She also attended Palm Sunday mass with six other children from the shelter the day before her first scheduled court appearance, and that offered some solace. No one knew how long La Salle's trial would last, and they tried not to bring the subject up with Sally, lest she get upset. The place she really wanted to be, after all, was home.

Thanks to an unexpected development, Sally's stay at the center didn't last long at all.

SEVENTEEN

A Guilty Plea

Frank La Salle wasn't allowed to travel from California to Camden by plane. Airline regulations at the time did not allow for passengers to be shackled, and Mitchell Cohen wasn't about to take any chances that the man would escape. "It is possible he could be a docile prisoner," Cohen remarked. "On the other hand, he could cause trouble."

The solution was to transport La Salle by train. Doing so would increase the travel time from hours to days, but on the train he could stay handcuffed to an officer for the entire duration. Marshall Thompson got stuck with being shackled to the prisoner for the cross-country trip, hardly a reward for all of his dogged investigative efforts. Wilfred Dube took the berth next door to Thompson and La Salle, staying as close as possible to the two men. (While it would have made sense for the two detectives to trade off being handcuffed to La Salle, I couldn't find any evidence that they did.)

Mitchell Cohen was at the train station to see La Salle and the detectives off. Before La Salle boarded, he asked Cohen why he and Sally Horner weren't getting on the same train. Cohen explained the two were due to fly later on in the day.

"Well, take good care of Sally," said La Salle.

"I'll take better care of her than you did," Cohen replied.

The train trip took two nights and two days. La Salle, Detective Thompson, and Detective Dube left San Francisco at 5:00 P.M. Pacific time on the *City of San Francisco*. Overnight the train passed through Sacramento, Salt Lake City, Cheyenne, Omaha, and Council Bluffs and reached Chicago early Saturday morning, where the trio changed trains to the New York–bound *General*. Thompson had no relief or privacy, shackled to the man he'd been chasing for nearly two years. Just as La Salle could not escape the law, so could the law not escape La Salle.

The *General* pulled into North Philadelphia Station at six minutes before seven in the morning on April 1. To the surprise of waiting reporters and photographers, the trio of men were not on board. To avoid the scrum, they'd gotten off at an earlier stop in Paoli, met there at 6:30 A.M. by Assistant Prosecutor William Cahill and Camden County Police Captain James Mulligan.

They took La Salle directly to the prosecutor's office. Then Thompson went home, no doubt relieved to be free of the man. Dube, Mulligan, and Cahill stuck around for Cohen's interrogation of La Salle, which lasted about four hours. At 1:00 P.M., La Salle was taken to the Camden County jail.

Mitchell Cohen told the press later on Sunday that he expected the case to go before the jury no earlier than June. Early on the morning of April 3, 1950, the day La Salle was due to be arraigned on the abduction and kidnapping charges, Cohen received a phone call from the county jail. The accused wanted to talk.

Cohen arrived at the jail at 9:45 A.M. and discovered La Salle by himself in a waiting room. He still lacked a lawyer—he hadn't been able to keep on Manuel Gomez because Gomez was not licensed to practice outside of California.

Cohen reminded La Salle of his right to an attorney. If he couldn't afford one, the court would appoint a lawyer for him.

"I don't need any counsel," La Salle replied. "I am guilty, and I am willing to go in and plead guilty. The sooner the better. I want to get it off my chest, and I want my time to commence to run."

When Cohen asked him why he wanted to plead guilty, La Salle said, "I want to avoid this girl [receiving] any further unfavorable publicity."

Cohen told him that court was already in session, and he could immediately enter his plea.

"Then I want to get it over with now," said La Salle.

Cohen left the Camden County jail and went straight to the courthouse for the arraignment. The courtroom was packed with onlookers. Sally Horner took her seat at the back next to a detective assigned to guard her. She wore a blue suit, pink blouse, straw hat, and patent leather Mary Jane shoes.

At ten minutes to noon, La Salle filed in, wearing a navy-blue suit, white collared shirt, and tie.

As Judge Rocco Palese entered, the room rose to attention. Like Mitchell Cohen, the judge had tangled with La Salle before. Palese, then a lawyer, had worked on Dorothy Dare's divorce petition against La Salle in 1944, even filling in for Dorothy's main lawyer, Bruce Wallace, at one of the hearings while La Salle was still serving time for statutory rape.

Palese, to the best of anyone's knowledge, never disclosed this prior association with the defendant. Perhaps he did not remember. Perhaps he didn't see a conflict because he had never engaged La Salle directly in court. Camden County's legal world was so small that defense attorneys became prosecutors who then became judges, everyone working with everyone else. What mattered was that, right now, Frank La Salle was in Judge Palese's courtroom.

When the gallery took their seats again, Palese called on Cohen to begin.

The prosecutor first outlined the story of Sally's kidnapping and confinement. How La Salle "persuaded and enticed her" to leave her mother in mid-June 1948, and told her that his repeated rapes of her were "natural." How Ruth Janisch "broke La Salle's spell" in San Jose, and Sally made the fateful phone call to the Panaros. How La Salle's long criminal record and his deviant behavior toward Sally made him, in Cohen's estimation, "a menace to society—a depraved man and a moral leper."

Cohen addressed both the judge and the crowd with his final words: "Mothers throughout the country will give a sigh of relief to know that a man of this type is safely in prison. That La Salle is somewhere safe, unable to harm anyone else."

Judge Palese asked Cohen if he had anything further. He did.

"If the Court please, at this time I propose to take pleas from this defendant, Frank La Salle, but before doing so I want the Court to know that I have discussed this case quite at length with this defendant, I have advised him of his right to be represented by counsel, and to have the benefit of the advice of counsel. I have warned him of the seriousness of these charges, and the length of the sentences which these charges impose.

"Being in Court is not an unfamiliar subject to this defendant and he understands, and he has told me so, the contents of both indictments. He understands the seriousness of them and the sentences, the minimum sentences, which the Court can impose, and he has advised me he does not desire to obtain counsel either on his own or appointed by the Court. I do, however, feel, in order that this defendant may properly understand the situation that it may be best for the Court to repeat to him his rights before I take the pleas."

Judge Palese then addressed Frank La Salle. "Mr. La Salle, you have just heard the prosecutor advise the Court that he has talked to you and explained to you the two indictments that have been returned against you by the Grand Jury of our County, and he has indicated to the Court that you do not desire to be represented by counsel and you desire to proceed in the matter without representation, is that correct?"

La Salle replied, "Yes."

"You understand the seriousness of the two indictments that have been returned against you by our Grand Jury?" Palese asked.

"Yes, sir."

"And that they carry with them rather serious sentences?"

Again, La Salle replied, "Yes, sir."

Judge Palese asked how La Salle would plead.

"Guilty," he said, in a voice barely audible.

The judge asked if there was anything more La Salle wished to say before he handed down the sentence.

La Salle said, his voice still weak: "I don't want any more publicity for the children." (Cohen later explained to reporters that La Salle was likely nervous and meant to say "child.")

Just like that, the proceeding was over. The whole matter took perhaps twenty minutes, ending just after noon. But not

before Palese decided upon a sentence for Frank La Salle. The judge
ordered Sally's abductor to spend no
less than thirty and no more than
thirty-five years in prison for the kidnapping charge. He would have to
serve at least three-quarters of the
full sentence before being eligible
for parole. Palese also added a two- to
three-year sentence for the original
abduction charge, as well as an additional two to three years for violating
his parole.

Two days later, just after noon on
April 5, La Salle began his sentence
at Trenton State Prison.

*Frank La Salle, after
pleading guilty.*

BECAUSE FRANK LA SALLE pleaded guilty, Sally did not get to
testify against him. In Mitchell Cohen's office after the hearing, she asked, once more, letting go of the earlier courtroom
stoicism and blinking back tears, when she could go home.
With the case finished, and Frank La Salle going to prison,
surely she could return to her mother right away?

Cohen sympathized with Sally, and told her so. There
seemed no reason to keep her in the state's custody when the
case was finished and La Salle incarcerated, but the wheels of
bureaucracy turned at their own pace, not his. Judge Palese
was the one who would have to decide when she could be released into Ella's custody again. Palese did so the very next day.

At noon on April 4, 1950, Cohen summoned Sally and Ella
to his office for what he later told the press was a "lengthy con-

ference" in which he delivered the news both of them wanted to hear the most. He also offered mother and daughter some advice. They were free, of course, to return to 944 Linden Street, but he thought it best they "went away from this area, changed their names and began life anew."

The extensive media coverage meant all of Camden, and much of Philadelphia and the surrounding towns, knew what had happened to Sally. Cohen worried the girl might be judged harshly for the forcible loss of her virtue, even if that reaction was in no way warranted. Cohen also urged Ella to seek the advice of the Reverend Alfred Jass, director of the Bureau of Catholic Charities, "in directing Sally's return to a normal life." Ella was a Protestant, but clergy was still clergy, and Sally's recent attendance at Catholic schools may have influenced Cohen's choice of religious advisor.

Sally and Ella got home at 1:45 P.M. Waiting reporters and photographers shouted questions and snapped pictures as they walked through the front door, Ella shielding her daughter. Ignoring the shouts, she shut the door firmly behind them.

From that afternoon on, the Horner women were private citizens. They were no longer at the mercy of the legal system, or the national press. The rest of the world could leave them alone.

In some fashion, it worked out that way, but their newfound calm did not last for long.

EIGHTEEN

When Nabokov (Really) Learned About Sally

Vladimir Nabokov spent the morning of March 22, 1950, much as he would every morning for the next month: bedridden and pain-plagued from the same neurological malady that had afflicted him a decade earlier, in the months leading up to his arrival in America. "I have followed your example and am in bed with a temperature above 102 degrees," Nabokov wrote Katharine White, his editor at the *New Yorker*, on March 24. "No bronchitis but grippe with me is invariably accompanied by the hideous pain of intercostal neuralgia."

White had also been ill and advised Nabokov to prize rest above work. Nabokov rested, but did not stop working. Just as he had written *The Enchanter* while bedridden a decade earlier, so now did he complete two late chapters of *Conclusive*

Evidence, the first version of the autobiography that became *Speak, Memory.* But as Nabokov told James Laughlin, his editor at New Directions, a month later, he did not "get back to normal conditions" for weeks. That summer, he and Véra did not travel across America to hunt butterflies, as they had done the previous year and on three earlier occasions. Not enough time, not nearly enough money, and too many deadlines loomed as his health slowly mended.

It's easy to imagine that, as he was laid up in bed at home in Ithaca with limited capacity to work, Nabokov picked up a copy of the local newspaper and came across the news of a kidnapped girl rescued in California after almost two years of cross-country captivity. It is not difficult to believe Nabokov, whom Véra described in their diary as being fascinated by true crime, paid avid attention from his sickbed as each day brought fresh news about Sally's rescue and Frank La Salle's crimes.

Here, in newspaper accounts of Sally Horner's plight, was a possible solution to a long-standing problem with the manuscript that would become *Lolita:* how to create the necessary scaffolding for all of the ideas rattling around in his mind, the decades of compulsion, and the games he wished to play with the reader.

Robert Roper, the author of *Nabokov in America,* was certainly convinced that Nabokov "read newspaper reports of a sensational crime" around the time of Sally's rescue. He told me, "I think reading about Sally was momentous for [Nabokov]. He was on the verge of abandoning his project when the March 1950 stories appeared, and it was as if the world were providing him with justification and template for writing his daring little sex novel. He cribbed so much from the story."

Yet there is no direct proof that Vladimir Nabokov learned of Sally Horner's abduction and rescue in March 1950. There was no story in the papers he was most likely to read—the *Cornell Daily Sun*, the college newspaper, or the *New York Times*. Similarly, there's no direct proof he glanced at the Camden or Philadelphia papers, the ones that carried the best details and the most vivid photos. Neither his archives at the New York Public Library nor those at the Library of Congress contain newspaper clippings about Sally. Any connection dances just outside the frame.

However, there is plenty of indirect proof that Nabokov knew about Sally Horner and her rescue. The circumstantial evidence is there in *Lolita*. And I believe he would never have fully realized the character of Dolores Haze without knowing of Sally's real-life plight.

LET'S FIRST CONSIDER HOW, roughly at the halfway point of *Lolita*, Humbert Humbert threatens Dolores into complying with him. He tells her that if he is arrested or if she reveals the true nature of their relationship, she "will be given a choice of varying dwelling places, all more or less the same, the correctional school, the reformatory, the juvenile detention home. . . ." Humbert's ultimatum echoes La Salle's repeated threats to Sally Horner, reported in the newspapers in March 1950, that if she did not do what he said, she would be bound for juvenile hall.

But earlier in the same scene, the comparison between Humbert and Frank La Salle is even more explicit: "Only the other day we read in the newspapers some bunkum about a middle-aged morals offender who pleaded guilty to the violation of the Mann Act and to transporting a nine-year-old girl across state lines for immoral purposes, whatever they are. Dolores darling! You are not nine but almost thirteen, and I would not advise you to consider yourself my cross-country slave. . . . I am your father, and I am speaking English, and I love you."

As Nabokov scholar Alexander Dolinin pointed out in his 2005 essay linking Sally Horner to *Lolita*, Nabokov fiddled with the case chronology. The cross-country journey in *Lolita* begins in 1947, an entire year before Sally Horner's abduction. At that time, Sally would have been nine going on ten, matching the age Humbert cites to his Lolita instead of the age she was when Frank La Salle abducted her. It is clear to Dolinin that "the legal formulae used by [Humbert Humbert] as well as his implying that he, in contrast to La Salle, is really Lolita's father, leave no doubt that the passage refers to the newspaper reports of 1950. . . ." In other words, the circumstantial evidence is right there in the text that Nabokov did, in fact, read about Sally Horner in March 1950, rather than retroactively inserting her story into *Lolita* several years after the fact.

To throw off the scent, or perhaps to amuse himself, Nabokov assigned details of La Salle to other characters. Dolores's eventual husband and the father of her child, Dick Schiller, is a mechanic. Meanwhile, Vivian Darkbloom—an anagram for Vladimir Nabokov—has a "hawk face," a phrase akin to the description of La Salle as a "hawk-faced man" in the March 1950 coverage of Sally's rescue. And as Dolinin underscored, references to Dolores's "Florentine hands" and "Florentine breasts" seem to point as much to Sally Horner's legal first name of Florence as they do to Botticelli.

Sally's captivity lasted twenty-one months, from June 1948 to March 1950. At the twenty-first month mark of their connection, Lolita and Humbert land at Beardsley, where Humbert realizes that he no longer has the same hold on the girl he once possessed. He worries Dolores has confided the true nature of her relationship with her "stepfather" to her school friend, Mona. And that in doing so, she might be cherishing "the stealthy thought . . . that perhaps after all Mona was right, and she, orphan Lo, could expose [Humbert] without getting penalized herself."

Dolores's potential confession to Mona echoes Sally's actual confession of her abuse at Frank La Salle's hands, first to the unnamed school friend, and later to Ruth Janisch. And just as Sally's escape comes about because of her long-distance phone call to her family, so, too, does Dolores make a mysterious phone call—immediately after fighting with Humbert—and then announces, "A great decision has been made." She doesn't flee him for another month, but the setup is already in place.

Then there is Humbert's aside in *Lolita*'s final chapter. He states that he would have given himself "at least thirty-five years for rape, and dismissed the rest of the charges." The exact sentence Frank La Salle received.

NINETEEN

Rebuilding a Life

S ally Horner was only two months past her eleventh birthday when Frank La Salle spirited her away from Camden. She returned home less than two weeks before turning thirteen on April 18. "When she went away she was a little girl," Ella murmured on the day she was finally reunited with her daughter. "Now she is practically a young lady." Sally had seen the country and how different so many other places were from Camden. She had been forced to grow up in the cruelest way possible, knowledge foisted upon her that could not be suppressed.

How the family marked her birthday isn't known, since no one, aside from Sally's niece, Diana, is alive to recall—and Diana was only twenty months old at the time. But a family outing to the Philadelphia Zoo, captured on a minute-long film clip shot by Sally's brother-in-law, Al Panaro, appears to provide a possible answer. It is the only known surviving footage of Sally.

In it, Sally seems dressed for spring, wearing the same outfit that she had on the plane from California, as well as to court the day that Frank La Salle pleaded guilty to her kidnapping. Her sister, Susan, has on a cream or white coat covering a pale blouse and dark skirt, while Diana is dressed in a pink two-piece suit.

Sally walks, shoulders hunched, beside Susan. At one point she pushes her niece in a white-handled stroller. She moves slowly, with hesitation, but it's not clear whether that's how she really moved or if the film clip was preserved at a slower speed.

In a close-up, Sally's face is angled to the left. Her expression is tentative, suggesting she still feels vulnerable out in public. That even though she is among her family, among those she loves, she isn't ready to let down her guard.

She does not look at the camera once.

THERE WERE OTHER PRESSING MATTERS as Sally Horner readjusted to life with her family, in Camden and elsewhere. She had been taken at the tail end of sixth grade; in the fall she would start eighth grade at Clara S. Burrough Junior High School, and was eager for what promised to be a fresh start. When she had gone to school during her captivity period, her energy was focused on surviving each day with Frank La Salle instead of dreaming about what she might want to be when she grew up. Now that Sally was free, she could think of what she wanted, for her own future. "She has a definite ambition," the San Jose detention center matron had said a few days after Sally's dramatic rescue. "She wants to be a doctor."

Ella, who had been out of work, needed to find a new job to support not only herself, but also a daughter who, through no fault of her own, was far closer to womanhood than any thirteen-year-old was supposed to be. Ella's repetition to the press of the phrase "whatever Sally has done, I can forgive her" points to her discomfort about the abuse Sally suffered, or even her lack of comprehension.

There was no vocabulary, in 1950, to describe the mechanism or the impact of Sally's victimization, where the violence was psychological manipulation, not necessarily brute force. Where the innocent-seeming facade of the father-daughter dynamic masked repeated rapes, unbeknownst to almost everyone around her. For Ella, who was struggling to pay the bills, put food on the table, and keep the lights on in the house, the details of Sally's captivity may have been too much to bear. As was the idea of starting over where no one knew what had happened to her. The stigma they knew must have seemed a better choice than the uncertainty of what they didn't know.

Taking Cohen's advice under consideration, Ella opted for a compromise: Sally would spend the summer of 1950 with the Panaros in Florence, while Ella remained in Camden. No one changed their names, and no one would discuss what happened to Sally for decades.

Over the summer of 1950, Sally Horner allowed herself to feel safe. She looked after Diana when Susan and Al Panaro had to work in the greenhouse, and sometimes Sally tended to the flowers and herbs as well. One family photo shows Sally in the greenhouse next to Susan, wearing dungarees, a white shirt, and a dark cardigan, her curly hair tousled around her face and chin, her mouth open as she is caught in mid-conversation with her sister.

*Sally Horner and her older sister, Susan Panaro,
in the family greenhouse.*

Other photographs from the same time suggest that living at the Panaros did Sally some good. One shows Sally standing by herself, clad in an elaborate pale-hued frock suitable for going to church or an afternoon social event. She's smiling at the camera, though her eyes carry remnants of the shyness she displayed while being filmed at the Philadelphia Zoo.

Sally's smile is wider in a second photo of her in a different fancy dress. Here she poses with a dark-haired young man wearing a suit at least two sizes too large. The boy is her apparent date, for a school dance or a church social. His name, and how the evening went, is lost to time—as is whether he was aware of what had happened to Sally.

By Sally's fourteenth birthday in April 1951, she looked like the typical American teenager of that period, the type to be wowed by Perry Como or Tony Bennett or Doris Day or other popular singers of the time. (In *Lolita,* Nabokov dutifully listed the soundtrack of Dolores and Humbert's road trip, including Eddie Fisher's "Wish You Were Here," Peggy Lee's "Forgive Me," and Tony Bennett's "Sleepless" and "Here in My Heart.")

One candid photo, likely taken by Al Panaro, hints at more complicated undercurrents in her than in a "bobby-soxer," as Sally was sometimes referred to in the press coverage of her rescue. She wears jeans again, as she did in the greenhouse photo with Susan, but now her shirt is dark, and her curly hair is pulled back. Her lipstick looks near-black in the black-and-white photo, which suggests she is wearing a ruby-red shade. The camera has captured Sally as she emerged from her bedroom, newspaper in her right hand, expression quizzical, as if she'd been interrupted while reading the funnies. She seems to be in need of sleep, caffeine, or a combination of the two.

Though Sally adopted a mask of good-natured resilience, Al recalled his sister-in-law drifting into melancholic moods. She would be in the moment, then gone. A light would shine, and

A candid shot of Sally holding a newspaper.

then flicker out. "She never said she was sad and depressed," Al told me in 2014, "but you knew something was wrong." The family discouraged discussion about her ordeal, and she almost never spoke of what happened with anyone. There were no heart-to-hearts. She underwent no psychological examinations; nor did she see a therapist. There was only Before, and After.

At Burrough Junior High, located on the corner of Haddon and Newton Avenues, Sally, once more, excelled on the academic side. Al recalled his sister-in-law being "very smart, an A student," and said that "it seemed like she knew a subject before it was taught." She graduated in June 1952 with honors.

Despite the photo of Sally with a date, her social life did not open up. She'd had trouble making friends before her abduction; afterward it became even more difficult. Classmates whispered and gossiped about her time with La Salle. Boys, emboldened and entitled, peppered her with unwanted remarks and propositions. As her classmate Carol Taylor—née Carol Starts—remembered, "they looked at her as a total whore." Emma DiRenzo, whom Sally knew as Emma Annibale, agreed. "She had a little bit of a rough time at first. Not everyone was very nice. I think some people didn't believe her."

It didn't matter to Sally's classmates that she had been abducted and raped. That she was not a virgin was enough to taint her. Nice girls were supposed to be pure until marriage. "No matter how you looked at it, she was a slut," Carol said. "That's the way it was in those days."

Carol met Sally in eighth-grade homeroom. Carol had street smarts; Sally did, too, but she wanted to close the door on how she got them, and escaped into the land of books. Carol lived two blocks away from the junior high while Sally had a longer daily walk of four to five blocks. Carol came from a large family—she was one of ten siblings, a far cry from Sally's smaller pool of immediate relatives. Carol had some other friends. Sally had no one but Carol, who didn't care a whit what anyone else thought of Sally. Carol said she was oblivious about Sally's supposedly sullied reputation, but it's as likely Carol chose not to behave the same way as her classmates, and not to judge Sally so harshly. Carol admired Sally's manners, her love of books, and sophisticated outlook. Sally admired Carol's freedom. She was as eager to be Carol's friend as Carol was to be hers.

Sally found refuge in the outdoors. She loved everything about being outside: the sun, swimming, and especially the Jersey Shore. As a little girl, before Frank La Salle kidnapped her, she'd spent many summer weekends at various seaside towns, like Wildwood and Cape May. After her rescue, the beach was a place where she could forget about cruel taunts and pervading despair. The Shore couldn't solve all of her problems, but at least it provided space for her to feel happiness.

In the summer of 1952, Sally was looking forward to starting Woodrow Wilson High School. At fifteen, she looked far older than her years. She wanted to make more friends and find a boyfriend.

Then, one weekend in the middle of August, she took another trip to Wildwood.

TWENTY

Lolita Progresses

The Nabokovs couldn't afford to road-trip across America during the summer of 1950, but the next summer Vladimir and Véra left Ithaca in June, at the end of Cornell's spring semester. Vladimir had turned in his grades for his European fiction class, and they gave up the lease on the house on East Seneca Street, their home the last three years, having found cheaper accommodations for the fall.

By the time Véra turned their aging Oldsmobile off U.S. Highway 36 at St. Francis, Kansas, on June 30, the pattern was set: hunt for butterflies for as many hours as a given day allowed, depending on their stamina and the weather. On rainier days—which dominated the trip—or when fatigue set in, usually in afternoons, Nabokov worked on the manuscript he was still calling *The Kingdom by the Sea.*

Véra and Nabokov chasing butterflies.

Nabokov worked in the passenger seat of the Oldsmobile, away from the noise coming through the motel room walls and insulated from the floods and storms that curtailed his exercises in lepidoptery. Dmitri, now seventeen, joined his parents in Telluride, Colorado—he was coming from Harvard, where he'd finished up his first year—and took over the driving duties, too. The family wended their way through the Rockies, Wyoming, and West Yellowstone, Montana, before returning to Ithaca at the end of August.

Weeks of butterfly-hunting in the Rockies, often shirtless with his chest exposed to the sun, had little immediate effect upon Nabokov's health. The accumulated exposure didn't cause any issue until he returned to Cornell, when a nasty case of sunstroke finally hit, confining him to bed for two solid weeks. "Silly situation . . . to be smitten by the insipid N.Y. sun on a dapper lawn," Nabokov noted in his diary. "High temperature, pain in the temples, insomnia and an incessant, brilliant but sterile turmoil of thoughts and fancies."

The Nabokovs changed their itinerary for the summer of 1952. They began their journey in Cambridge, Massachusetts, rather than Ithaca, because Vladimir had taken up a teaching post at Harvard for the spring semester (he was on sabbatical from Cornell). A mitigating factor in moving back to the Cambridge area was to be closer to Dmitri, continuing his studies at Harvard.

Vladimir, Véra, and Dmitri, driving the same Oldsmobile as in earlier years, landed in Laramie, Wyoming, at the end of June, about ten days after departing Cambridge. They stayed in the state hunting butterflies all along the Continental Divide. They traveled through Medicine Bow National Forest ("using the abominable local road") to Riverside in time for the Fourth of July (where "some noisy festival is underway") and, by early August, arrived in Afton.

All the while Nabokov continued to scribble down notes on index card after index card, adding to the novel that had bedeviled him for so long. He had spent the previous year sharpening his observations of quotidian matters. He noted down all sorts of minutiae, the better to portray the American prepubescent girl at the heart of his novel with greater accuracy. Nabokov recorded heights and weights, average age of first menstruation, attitude changes, even the "proper method of inserting an enema tip into a rectum."

He also jotted down teen magazine slang—which is why phrases like "It's a sketch" or "She was loads of fun" appear in Lolita and sound right, not tin-eared. To create the character of Miss Pratt, the Beardsley school head, Nabokov interviewed a real school principal under the guise of having a (fictional) daughter who wanted to enroll.

But he did not make as much progress on Lolita as he wished while wandering along the Continental Divide. The academic year had exhausted Nabokov more than he realized. He saved most of his energy to scour for blues, including a successful sighting of Vanessa cardui. In due course it was time for the Nabokovs to return east. Dmitri had gone back to Cambridge earlier, leaving his parents to travel by themselves along two-lane highways. The couple likely needed two weeks to make the 1,850-plus-mile trip back to Ithaca. They reached the town, and another new rental home, on September 1, 1952.

By that time, Nabokov had read a new story about Sally Horner, one that would change the direction of *Lolita* so much it's surprising to think the novel could have existed without it.

TWENTY-ONE

Weekend in Wildwood

Carol Starts, Sally Horner's best friend, summer of 1952.

Carol Taylor no longer remembers why she and Sally decided to go down to Wildwood that summer weekend in 1952. It was mid-August in Camden, a time of relentless heat and humidity. Nobody had air-conditioning, and heading to the Jersey Shore was an easy way to find some relief.

Carol and Sally were both working summer jobs as waitresses at the Sun Ray drugstore in nearby Haddonfield. They were best friends. They were fifteen. They were just a few weeks away from their first freshman class at Woodrow Wilson High. Why not head down to Wildwood for a quick getaway?

The girls saved up their pennies for bus fare and headed south on Friday, August 15, an hour-and-a-half-long ride covering just over eighty-six miles. They got there in the late afternoon. Wildwood bustled with the energy of all the young people making similar weekend pilgrimages for sun, sand, beach, and nightlife. Carol and Sally also carried fake identification cards saying they were twenty-one years old. This would lead to confusion later on.

Sally and Carol weren't drinkers. Sally did not touch alcohol at all, while Carol only occasionally sipped some beer or wine. The fake IDs weren't about boozing. If you wanted to dance, you had to go to clubs like the Bamboo Room, the Riptide, or the Bolero, and you needed to be over the age of twenty-one to get in. Every other Camden high school kid had a fake ID. Plus it was so easy: go to City Hall, get a card-sized version of your birth certificate, adjust the birth date, bleach it out then dye it green with vegetable food coloring, get it laminated, and voilà, a genuine-looking false identification card.

Sally and Carol hit the beach, and after that danced the night away. Then, on Saturday, the friends' plans split apart. For that was when Eddie met Sally.

EDWARD JOHN BAKER drove down to Wildwood nearly every summer weekend in 1952. When Sally Horner met him, he must have seemed like a boy for whom fun was something not just to be had, but to be lived.

In photographs from his high school yearbooks and local newspapers, Baker's eyes dance with merriment. It's there even in the classic high school graduation head shot: while other classmates try for seriousness belying their years, Baker, his dark hair tousled and his eyebrows raised, makes the thought of putting away childish things seem eminently unreasonable. The sparkle in his eye lurks in group photos of the many school musical groups he played with, from the senior jazz band to the string orchestra to the treble clef quartet. In

Edward Baker's high school graduation photo, 1950.

all those shots, Baker holds his trusty soprano saxophone like the proverbial piper ready to entice those eager to follow him wherever he may go.

Photographs lie, of course. They are but millisecond-long glimpses of a more complicated set of feelings, emotions, interactions. One should be careful of reading too much into them. But photos are all that remain, since Baker—Eddie in his youth—isn't around anymore to say what was in his mind at the time. He died in 2014, age eighty-two, still living in his hometown of Vineland, New Jersey.

What is clear from those photos of Baker is why Sally found him attractive at a time when she was finally ready to feel such things. Edward Baker was tall, dark, and twenty to Sally's mature-not-by-choice fifteen. She didn't tell Baker her real age; she said she was seventeen, and Baker said later he "thought she probably was. She looked it." Carol said that Sally was "bananas" for him as soon as she saw him.

Sally had confided in Carol throughout their yearlong friendship, and especially over that summer, about her loneliness and longing for a boyfriend. How that seemed impossible in Camden, where too many people knew about her kidnapping. Where she was viciously mocked by boys and girls alike. Branded a slut. Shunned.

Baker was a tonic to all that. Just the right amount of older, taller, handsomer. She wasn't about to correct him if he believed she was seventeen, or tell him that the new school she was about to start in the fall was Woodrow Wilson High. Perhaps she hoped Baker could pry her away from the darkness. Or perhaps Sally merely wanted a weekend diversion.

After meeting at the beach on Saturday, they spent the afternoon and evening together, and on Sunday morning they went to church. "She impressed me as a darn nice girl," said Baker. "She was good-looking, reserved . . . and was apparently a church-goer."

If something more happened between Saturday night and Sunday morning, Sally did not confide in Carol. Sally did, however, ask a gigantic favor from her best friend after she and Eddie returned from church: Would Carol be okay heading back to Camden on her own? If she was, Sally would go with Baker in his glossy black Ford sedan to his hometown of Vineland and catch a bus from there.

"She really, really, wanted to go home with him," recalled Carol. "She thought he was so nice."

Carol said sure, she didn't mind. She had no reason to get in the way of her best friend's infatuation. Baker seemed all right, not someone who would do Sally harm. Besides, other friends of Carol's were in Wildwood that weekend, too, and they had room for Carol in their car.

The ride home to Camden was tranquil for Carol. The following morning would be nothing of the sort.

ED BAKER PULLED onto the highway with Sally Horner in the passenger seat. Her spirits must have been high as they began the trip to Vineland. They'd spent all of Sunday together, just like they had on Saturday. She was dead gone on him, and he seemed to feel the same way about her. That evening, she'd met Eddie for supper, after which they walked along Wildwood's bustling boardwalk. Away from the carnival barkers and the screaming children, they spied a free bench and sat down. They talked and talked, and perhaps even kissed, and then headed for his car well after dark. She didn't want to say goodbye just yet.

The trip from Wildwood to Vineland normally took about forty minutes. Perhaps, given that it was after eleven o'clock at night, Ed Baker wanted to drive a little faster. Perhaps they were rushing to get Sally to the station before the last bus left for Camden. Or maybe their plans were not so innocent, and they didn't want anyone else to know.

As the clock neared midnight, Ed Baker and Sally Horner were seventeen miles north of Wildwood. A car approached from the opposite direction of the two-lane highway, and Baker shifted his headlights to low beam. He kept both hands on the steering wheel to hold the car in the center of the road. Through the glare, Baker caught a glimpse of something off to the side, but not fast enough to avoid a collision.

Sally never felt a thing.

JUST AFTER MIDNIGHT on Monday, August 18, 1952, the New Jersey State Police arrived on the scene of a four-vehicle highway accident on lower Woodbine Road where the county line divided North Dennis from Woodbine. (The entire stretch is now part of Highway 78.) Baker had barreled into the back of a parked truck, owned by Jacob Benson, which proceeded to crash into another parked truck, owned by John Rifkin. The impact caused Rifkin's truck to be thrown into the highway, where it was hit again by the car directly behind Baker's Ford.

State Trooper Paul Heilfurth told the *Wildwood Leader* that if the multi-car crash had happened three minutes later, it "would probably have been more serious." That's because Benson's truck was about to be towed by Rifkin's truck, and both men were safely away from the vehicles when Baker plowed into them.

Those three extra minutes saved the men's lives. Baker broke his left knee, needed fifteen stitches to close a gash on his right arm, and was cut up and bruised.

The car crash killed Sally Horner instantly.

Rescue crews took more than two hours to free Sally's body from the wreckage. Her head had been crushed by the truck's tailgate, which had come through the windshield when the vehicles collided. Police discovered the fake ID that claimed Sally was twenty-one. Initial news reports misreported her age as a result. Once they realized who she was and that she had been in the news before, Sally's true age emerged.

The death certificate, issued by Cape May County three days later, listed the cause of Sally's death as a fractured skull from a blow to the right side of her head. She'd broken her neck; other mortal injuries included a crushed chest and internal injuries, as well as a right leg fracture above the knee. The coroner didn't bother with an autopsy.

The damage to her face was so severe that the state police felt Ella would be too traumatized to identify her daughter. Instead, Al Panaro went to the morgue. "The only way I knew it was Sally," he said, "was because she had a scar on her leg. I couldn't tell from her face."

CAROL STARTS WAS WOKEN UP on the morning of August 18 by the sound of her mother yelling, "There's someone on the phone for you!" There was only one phone, located in the living room. Carol got herself up and rushed out to take the call. The person on the line sounded official, like a policeman or a detective.

The caller wanted to know if Carol had been with Sally Horner the previous night.

"Yes, I was," Carol replied.

"And you're aware who she was with?"

"Yes I am. Why are you asking me this?"

Carol could not grasp what the caller wanted. Without waiting for an answer, she hung up the phone, then picked it up again and dialed Sally's number. Ella answered.

"Hi, Mrs. Horner. Where's Sally? Is she up yet?"

Ella began to sob. Then she told Carol that Sally had died in a car accident the night before.

Things grew strange for Carol. She did not react right away to the death of her best friend. She got dressed, left the house, and went straight to the movie theater. "I don't know what I saw. I don't know what outfit I wore. But when people wanted to talk to me, I went to the movies." Later she understood she had gone into shock.

When Carol returned home, Ella called again. She explained in greater detail what had happened to Sally, describing the injuries she'd sustained, or at least what Ella knew of them. Only after hanging up did Carol feel the loss of her friend. "I cried and cried and cried."

Carol also couldn't bring herself to ask Ella about a more mundane matter. She and Sally had borrowed each other's favorite dresses in Wildwood—and Carol's blue frock was packed in Sally's bag. It wouldn't have been right to ask Ella about what happened to her dress, so Carol didn't.

But Carol told Ella she'd met the boy who'd taken Sally on her final, fatal journey.

TWENTY-TWO

The Note Card

On the morning of August 19, 1952, as he and Véra were about to begin the long drive back to Ithaca, Vladimir Nabokov opened up a newspaper somewhere near Afton, Wyoming, and chanced upon an Associated Press story. Perhaps the newspaper Nabokov read was the *New York Times,* which carried the wire report of Sally Horner's death on page twelve of their early edition. Maybe it was a local daily, which splashed the sensational news on or near the front page. Wherever Nabokov read the report, he took notes on one of his ninety-four surviving *Lolita* index cards.

The handwritten card reads as follows:

20.viii.52

Woodbine, N.J. –

> Sally Horner, 15-year-old Camden, N.J. Girl who spent
> 21 months as the captive of a middle-aged morals offender
> a few years ago, was killed in a highway mishap early
> Monday . . . Sally vanished from her Camden home in
> 1948 and wasn't heard from again until 1950 when she
> told a hararing [sic] story of spending 21 months as the
> cross-country slave of Frank La Salle, 52.
>
> LaSalle [sic], a mechanic, was arrested in San Jose,
> Cal . . . he pleaded guilty to (two) charges of kidnaping
> and was sentenced to 30 to 35 years in prison. He was
> branded a "moral leper" by the sentencing judge.

Here, in this note card, is proof that Nabokov knew of the Sally Horner case. It is proof that her story captured his attention and that her real-life ordeal was inspiration for Dolores Haze's fictional plight. Less clear is whether the wire report Nabokov read in August 1952 was the first time he had heard of the girl, or if he was, like all who had read the news stories in March and April 1950, stunned to realize that she'd only lived two more years after her rescue.

The note card, written on front and back, included a number of strikethroughs that ended up in the text of *Lolita* itself. Nabokov crossed out "middle-aged morals offender" and "cross-country slave," both phrases that serve as Humbert Humbert's justification to Lolita that the "bunkum" they read in the newspapers has no relation to their "father-daughter" relationship. Misspellings dotted the note card. The most notable is "harrowing," which Nabokov tortured into some alternate, Russified version of the word.

At the top of the note card Nabokov wrote: "in Ench. H. revisited? . . . . in the newspaper?" As Alexander Dolinin explained, Nabokov was referring to "a scene (Chapter 26, Part II) in

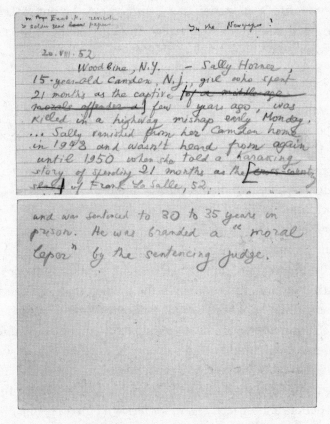

AP story of Sally Horner's death transcribed onto
a note card by Vladimir Nabokov.

which Humbert revisits Briceland and in a library browses
through a 'coffin-black volume' with old files of the local *Ga-*
zette for August 1947. Humbert is looking for a printed picture of himself 'as a younger brute' on his 'dark way to Lolita's
bed' in the Enchanted Hunters hotel, and Nabokov evidently
thought of making him come across a report of Sally Horner's
death in what the narrator aptly calls the 'book of doom.'"

Nabokov decided against this approach. Instead, he seeded

Sally Horner's abduction story throughout the entire *Lolita* narrative, making it a tantalizing thread for readers to discover on their own—though the vast majority never did.

AN ALTERNATE THEORY of the ending of *Lolita* pops up in Nabokovian circles from time to time. It contends that Dolores Haze, rather than meeting and marrying Dick Schiller, becoming pregnant, and then dying in childbirth before she is eighteen, actually died at the age of fourteen and a half. Her short, tragic adult life is in fact Humbert Humbert's delusion, a projected fantasy in order to create some sort of romanticized ending for the girl he defiled.

In this version, rather than bearing responsibility for her death, Humbert can indulge in the illusion that—at least for a short time—Dolores found her way to a kind of happiness. By extension, he can remold their rapist-victim power dynamic into real love. Humbert can convince himself he did not want Dolores because she fit the nymphet type born out of his childhood obsession with Annabel Leigh, but that he pursued the girl out of some special regard for her as a human being.

If that theory is true—Nabokov certainly never confirmed or denied—Humbert's final visit to Ramsdale carries an extra sharpness. Just before Nabokov invokes Sally Horner and Frank La Salle in a parenthetical aside, the reliable means through which he conveys true meaning to the reader, he has Humbert Humbert walking through Ramsdale, reminiscing about his first, fateful glimpse of Dolores Haze. Humbert strolls by his old house and spies a "For Sale" sign with a black velvet hair ribbon attached. Just then, "a golden-skinned, brown-haired nymphet of nine or ten" passes him, looking at him with "wild fascination in her large blue-black eyes."

She could be a composite of Lolita and Sally, her eyes the same color as Sally's, more or less. Humbert says, "I said something pleasant to her, meaning no harm, an old-world compliment, what nice eyes you have, but she retreated in haste and the music stopped abruptly, and a violent-looking dark man, glistening with sweat, came out and glared at me. I was on the point of identifying myself when, with a pang of dream-embarrassment, I became aware of my mud-caked dungarees, my filthy and torn sweater, my bristly chin, my bum's bloodshot eyes."

Here, so late in *Lolita*, Humbert has his moment of reckoning. He understands, briefly, "what he might really look like in the eyes of his eternal jury: children and their protectors." The glib charm, all of the smooth veneer, is stripped away in an instant. Humbert reveals himself as the monster he knows he is. And by killing Clare Quilty for taking Dolores away from him—in his mind, taking away what was rightfully his—Humbert Humbert loses his last vestige of morality.

Dolinin takes a charitable view of Nabokov's treatment of Sally Horner in *Lolita*, claiming that the number of references, including the architecture of the novel's second half, does not obscure the real girl. Rather, he writes, "[Nabokov] wanted us to remember and pity the poor girl whose stolen childhood and untimely death helped to give birth to his (not Humbert Humbert's) Lolita—the genuine heroine of the novel hidden behind the narrator's self-indulgent verbosity."

This sense of pity Dolinin speaks of emerges in Humbert's final meeting with Dolores. She is married, pregnant, and seventeen, with "adult, rope-veined narrow hands." She has aged out of his perverse desires, and he finally understands, through the use of the parenthetical, how much he defiled and violated her, how much damage he has caused:

". . . in the infinite run it does not matter a jot that a North American girl-child named Dolores Haze had been deprived of her childhood by a maniac, unless this can be proven (and if it can, then life is a joke), I see nothing for the treatment of my misery but the melancholy . . . palliative of articulate art."

Humbert's epiphany is in keeping with Véra's diary note only days after the American publication of *Lolita* in 1958. She was ecstatic about the largely positive press and fast sales of the novel, but was unnerved by what critics weren't saying. "I wish, though, somebody would notice the tender description of the child's helplessness, her pathetic dependence on monstrous HH, and her heartrending courage all along."

It is to Nabokov's credit that something of the true character of Dolores—her messy, complicated, childish self—emerges out of the haze of his narrator's perverse pedestal-placing. She is no "charming brat lifted from an ordinary existence only by the special brand of love." She excels at tennis; she is free with sharp comebacks ("You talk like a book, *Dad*"); and when she seizes the opportunity to break away from Humbert and run off with Clare Quilty, she does so in order to survive. Any fate is better than staying with her stepfather.

Never mind that she will, later, run from Quilty's desire to embroil her in pornography with multiple people. Never mind that Dolores will "settle" for Dick Schiller and a life of domesticity and motherhood that is, sadly, cut short. She still has the freedom and the autonomy to make these choices for herself, a freedom she never had while under Humbert Humbert's power.

These choices are likely why Véra rated Dolores so highly in the diary entry, and why Nabokov himself ranked Lolita second (after Pnin) of all the characters he ever created that he admired most as a person.

TWENTY-THREE

"A Darn Nice Girl"

On August 21, 1952, three days after the car accident that killed Sally Horner, the *Vineland Daily Journal* published a front-page interview with Edward Baker. He said he was "bewildered by publicity" over Sally's death. "I'd never met Sally before. She didn't tell me if she had ever been to Wildwood before, but I got the impression this was probably the first time she'd ever visited the place." Baker said he frequented the resort town "just about every weekend." On Friday, August 15, he left early from Kimble's, the glass plant where he worked as an apprentice machinist. He met Sally the next day, as well as "a whole bunch of other fellows and girls I met down there . . . Sally and I hung around with them most of the time."

He insisted news accounts of the car accident were wrong. "The fellow who owned the truck I hit [Benson] said he was on the shoulder of the road. But I certainly wasn't on the shoulder, and my skidmarks will prove it. The fellow behind me, even with the benefit of my lights, didn't see the truck, and he crashed into it also." Baker said that what saved his life was the fact that he had both of his hands on the steering wheel, which broke in the collision.

Three days of coast-to-coast news stories had rattled Baker's nerves, and he wanted to set the record straight about what happened between him and Sally. "She seemed like a nice girl. Some of the stories that followed the accident sounded as though we were making a sinful weekend of it. We didn't do anything wrong. . . . We weren't 'fooling around' in the car, or anything. If we had been, she probably wouldn't have been killed and I might have been."

Baker was even more flabbergasted by the revelations of Sally's past ordeal. "Nobody had any idea this girl was the one who had been kidnapped four years ago. How should we remember?" Never mind that Sally's rescue was reported nationwide, as well as on the front page of the *Daily Journal*.

He was still grappling with how young Sally really was. "She told me she was 17 years old. She may have had a birth certificate with her saying she was 21, but I never saw it. Who asks to see birth certificates when you go out with a girl?"

The *Daily Journal* also spoke to Baker's mother, Marie Young. She'd received a call from her son not long after the accident. "He said he wished it was him that was killed instead of that innocent girl. He was pretty broken up." He'd told his mother how nice a girl she was, and how he admired her commitment to going to church on Sundays, even down in Wildwood. He would never get past that "she got killed because she wanted him to take her into Vineland."

Both Baker and his mother had added reasons to defend

themselves. After he was treated at Burdette Tomlin Hospital in Cape May for the injuries he sustained in the car accident, police arrested and charged him with reckless homicide. Baker was freed on a thousand dollars' bail—his stepfather, James Young, put up the money—on August 20. A news account sympathetic to Baker, stressing his lack of culpability in the accident, might help his case.

But a strike against him was that the crash leading to Sally's death was not Baker's first car accident. Only the year before, Baker was driving the car, which belonged to his mother, Marie Young, in Newfield, four miles north of Vineland. He hit another car while running a red light. Then, too, Baker's injuries were not life-threatening. Neither were those of Marie, sitting in the passenger seat.

SALLY HORNER'S FUNERAL was held on August 22, four days after her death. More than three hundred people crowded into the Frank J. Leonard Funeral Home at 1451 Broadway to pay their respects. Many floral arrangements sent by well-wishers flanked Sally's casket.

The burial was a more private affair. Only a handful of family members, including Ella, Susan, Al, and some aunts and cousins, drove out to Emleys Hill Cemetery in Cream Ridge, where Sally's remains were interred in the Goff family plot.

For Carol Starts, the funeral was awful. She sat by herself in a corner pew. Ella and Susan requested the casket be open at first, for those who wished to pay their last respects to Sally. "I wanted to see her so badly. Then I did, and it nearly broke me in half," Carol recalled. When she could no longer stand the proceedings, Carol fled the service and went home.

Carol stayed away from school for an entire week after

Sally's death. "I couldn't handle it. This was the most heavy-duty thing I had ever gone through." Carol's first experience of deep loss would mark her for the rest of her life. As she grew older and friends began to die, Carol tended to grieve in an open and wild manner that puzzled those around her. "I would hear, 'but they were just a friend.' I would hear that about Sally. That we should be moving right along. I wasn't willing to move right along. I wanted to grieve. And when I finally came out of shock, I did."

FRANK LA SALLE made his presence known to Sally Horner's family one final time. On the morning of her funeral, they discovered he had sent a spray of flowers. The Panaros insisted they not be displayed.

THE FIRST COURT HEARING stemming from the accident that killed Sally Horner took place on Tuesday, August 26. It lasted more than two and a half hours. A full record of the proceeding does not exist, but the surviving court docket reported that Baker pleaded not guilty to a count of careless driving and that Judge Thomas Sears found him not guilty of the charge.

New Jersey state law enforcement was not about to let Baker go so easily, and what followed was a complicated series of charges, court hearings, and verdicts. Prosecutors even charged Baker for "operating a car with illegal equipment"—specifically, unapproved headlight shields. Police told the *Vineland Daily Journal* that Baker's headlights "were partially obscured by a device he had purchased and attached to other lights."

The most serious charge came from the Cape May grand jury. On September 3, 1952, they indicted Baker for the "reckless killing by auto" of Sally Horner, stating that he had acted "carelessly and heedlessly, in wilful and wanton disregard of the rights and safety of others . . . and against the peace of this State, the Government, and dignity of the same."

The following week, on September 10, Baker pleaded not guilty to that charge in front of Judge Harry Tanenbaum. Carol was called to testify to her friendship with Sally, as well as the whole business with the fake identification cards. Her memory of the experience was hazy, but she had vivid recall of Baker's attitude.

"He was very arrogant," Carol told me. "He would make these weird remarks, like how the courtroom was only thirty feet long instead of being a hundred feet as it was supposed to. I didn't understand what he meant and still don't." She was still so worked up about his comment about the courtroom size that she mentioned it to me three times in one conversation. It demonstrated, to her, Baker's inability to take the hearing seriously: "He snickered a great deal. Acting stupid." Her reaction to him was visceral: "I hated him because he was driving and had the accident that killed my best friend."

Perhaps Carol was also upset by the court's decision, delayed until January 15, 1953. Judge Tenenbaum threw out the charge against Baker for reckless killing by auto—the sketchy court records did not give a reason—and also found Baker not guilty of the unapproved headlight shields count.

But Baker's legal troubles were not done. He faced a cluster of civil actions, too. As with the criminal court proceedings, the surviving civil court documents are light on detail and full of unresolved gaps. But both the *Cape May County Gazette* and the Camden *Courier-Post* reported important details on the complicated joint lawsuits.

All five complaints were heard together the week of May 21,

1953. Dominick Caprioni, who owned the car right behind Baker's on the night Sally Horner died, sued Jacob Benson, owner of the parked truck that both Baker and Caprioni crashed into. Caprioni also sued Baker and his mother, Marie Young, since she owned the Ford Baker was driving, seeking $13,300 in damages in his lawsuits. Benson sued Baker and Caprioni without asking for any money, while Baker and Young in turn sued Benson for $52,500. Lastly—and the civil action that matters most—Ella Horner sued Baker, Young, and Benson for $50,000.

The byzantine nature of the lawsuits, heard by Superior Court judge Elmer B. Woods, may explain why there was a mistrial on the first day, after someone observed a juror talking to one of the witnesses during the noon recess. A new hearing lasted two days before ending in an abrupt settlement on May 28, 1953. It's not clear how much each of the plaintiffs (some of whom were also doubling as defendants) received.

The cases against Baker were over, but Cape May County did not officially close the books until June 30, 1954. Written beside his name on the ledger for that day was "nolle pros"—they declined to prosecute. At last the car crash that killed Sally Horner, in the eyes of the criminal justice system, had been ruled a tragic accident.

TWENTY-FOUR

La Salle in Prison

Frank La Salle's lengthy prison sentence meant Sally Horner's family could push her abductor to the back of their minds. The New Jersey court systems, however, weren't so lucky. La Salle may have pleaded guilty and shrugged off his right to an attorney, but he still thought he could find a way out of prison, banking on an outlandish fantasy about the life he'd led with Sally. Unlike Humbert Humbert, who tried to attach some grander meaning to his delusions of exemplary parenthood, La Salle came up with a rationale that was crude, obvious, and disappointingly banal.

He applied for, and received, a writ of habeas corpus from the Mercer County Court (Trenton State Prison fell under their jurisdiction). The gist of La Salle's first appeal was that he'd never waived his right to extradition from California and was thus brought to Camden against his will. In lengthy testimony at the Mercer County Courthouse on September 24, 1951, La Salle also claimed he "did not plead guilty before Judge Palese in Camden" and was thus "deprived of his liberty without due process of law."

While the transcript of the proceeding is lost, I discovered the outcome of La Salle's habeas hearing in a motion filed after the fact by Camden County prosecutor Mitchell Cohen: La Salle perjured himself on the stand by denying he had pleaded guilty to the abduction and kidnapping charges when, in fact, he made that plea in open court in April 1950.

Future New Jersey governor and chief justice of the State Supreme Court Richard Hughes presided over the case as a county court judge. Hughes was so incensed by La Salle's lies on the stand that he told the prisoner: "I hardly think you should be able to finish your first [prison] sentence, but in case you should, I am imposing another. You have attempted to subvert and obstruct the proper administration of law by an application for a writ [of habeas corpus] with absolutely no foundation." Hughes ordered La Salle to serve an additional thirty days in the Mercer County Jail.

Judge Hughes's ire toward La Salle was compounded by the recent uptick of prisoners acting as their own jailhouse lawyers who showed no compunction about lying in their filings. "The courts always lend a willing ear where the deprivation of rights are concerned," Hughes wrote, "but of late there has been a great abuse of this privilege by prisoners who lie when they file sworn statements. It must be broken up."

The contempt of court finding did not deter La Salle from his appeals. He kept on, in a lengthy series of motions and affidavits filed in waves between 1952 and 1955. They are the only surviving evidence of La Salle's state of mind during and after Sally's abduction, and they are remarkable proof of a false narrative. The authority and control he radiated when he had Sally in his power, his ability to manipulate others into believing in his act, vanished upon his arrest in March 1950. Now he exuded desperation and falsity.

In his appeals, La Salle refused to acknowledge that Sally was not, and had never been, his daughter. He referred to her repeatedly as "Natural Daughter Florence Horner La Salle," apparently having convinced himself, through the selective reading of legal precedents, that "a father cannot be convicted of kidnaping his own child."

He spun a shambling yarn about living in Camden, but "not with his family," in January 1948, "doing what he thought right by giving money in sufficient sums to his former common-law wife for the care and maintenance of his [daughter]" and seeing Sally "on the streets by herself as late as 12 AM midnight," at which point he would "make her go home and give her money." La Salle neglected to mention his January 15, 1948, parole on the statutory rape charges, or that he never knew, let alone had any domestic relationship with, Ella Horner. And he most certainly never saw Sally out at midnight; nor did he give her money and tell her to go home.

La Salle justified his kidnapping of Sally, both in his appeals and, later, in person, to his actual daughter, Madeline, on the grounds that he was saving Sally from a mother who was "always out with some man or [was home] in bed," or by falsely quoting Sally saying that her mother "does not care what becomes of me. She seems to hate me, and never buys any clothing or take [sic] care of me and is never at home."

He embellished these poorly written fantasies of devoted fatherhood to Madeline, a daughter he never saw grow up, by describing trips to Philadelphia "to see his other daughter by his legal wife who he was at the time separated from, but there was nobody at home." (Dorothy Dare, of course, had filed for divorce from La Salle in 1943 after he was arrested on the statutory rape charges.)

La Salle attested, time and again, to having "sworn proof" that Sally was his daughter, but of course he could never deliver the goods. He even reproved the media for publishing Sally's name after she was rescued in San Jose on the grounds of "a statute against such publicity for a child." He claimed his quick guilty plea resulted from being afraid of "MOB VIOLENCE" (the capitalization is La Salle's) and also claimed that the prosecutor, Cohen, "told the defendant there was no use of his trying to get an attorney as no attorney could do any good."

La Salle's appeal documents include purported affidavits that bolstered his claims of loving fatherhood. If the documents are real, they show how many missed opportunities there were for one of Sally's neighbors to see past the facade of amiable father-daughter interaction to the horrifying reality. If the documents are forgeries, they amplify the grimy, sordid truth of Sally's abuse: she was under the power of a man so determined to present himself as a well-adjusted human and bury the depth of his crimes that he lied, above all, to himself.

MOST OF THE STATEMENTS attributed to Sally Horner and Frank La Salle's neighbors in Dallas come from affidavits included with La Salle's appellate brief in 1954. After reading a copy of the statement his mother, Nelrose Pfeil, purportedly gave about Sally, Tom Pfeil denied she'd ever said anything of the sort. "It was definitely not the way my mother talked. Not her wording or verbiage, anyhow," he told me. Frequent misspellings in the alleged affidavit also gave Tom pause: "My mother was a good speller," he said. "She was a secretary for three attorneys just out of college. She may not have had a legal mind, but she was precise."

Tom Pfeil scoffed at his mother's supposed statement that Sally spent "many hours a day" at the family home. For one thing, none of the Pfeils were home much. Charles and Nelrose also owned and operated a lumberyard along with the trailer park. They worked sixteen- to eighteen-hour days during the week, and the boys started working at the yard in their teens, considering it a victory when they negotiated a weekly half day off. When Tom joined the marines out of high school and began boot camp, he told me he thought, "Man, I've been down this road already."

"My mother was a very strong woman," said Tom. "She wasn't but five foot one but she could thread iron into a water pipe. If a water line froze or broke, [Nelrose] would put it back in the drain. She wasn't mean, but running a trailer park wasn't running a bridal store." She had such a solid work ethic that she worked at the yard most every day until four months before she died in 2001, at the age of eighty-four. She and Charles had an ironclad rule at the trailer park: no socializing with the tenants. "She collected the rent, and not much other than that," said Tom. "That was one of the things they decided because they got hurt. A couple of times, I befriended some people I shouldn't have and, unfortunately, they took my parents to the cleaners. As soon as you become a friend, it becomes, 'Oh, I can't pay this week.'" As a result, the Pfeils kept their interactions with the trailer park residents to the bare minimum. "Sally would *not* be in the house several times a day," Tom emphasized.

His recollections, however, jibe with what his mother's supposed affidavit said about La Salle spoiling Sally. "She wasn't laying [*sic*] on the ground kicking. . . . When she asked [La Salle] for something, she got it," said Tom. "She had nice clothes. She wasn't running around in rags. Certainly she was not mistreated. That's why everybody was surprised. We thought it was a dad's adoration of a daughter and how nice that was."

One curious anomaly with Nelrose Pfeil's alleged affidavit was that it included the family's new address, 2240 Lawndale Avenue. If the affidavit was a fabrication, as her son insisted, how would Frank La Salle have known where the Pfeils had moved after they stopped living at the Commerce Street trailer park?

Tom Pfeil was adamant that his mother was never in touch with La Salle while he was in prison. "For my mother to have anything to do with an affidavit or anything . . . is about as farfetched as if she came back from Mars."

FRANK LA SALLE also wrote letters while incarcerated at Trenton State Prison, just as he had during earlier prison stints. According to some of Ruth Janisch's children, he wrote their mother on a number of occasions. Both Vanessa Janisch, who was not born until after Sally was rescued, and her older sister Rachel* remembered seeing La Salle's letters bundled up and tucked in one of the many scrapbooks—at least one chiefly devoted to media coverage of Sally's rescue—that Ruth kept of her life.

As of this writing, I have not been able to see Ruth's scrapbooks for myself. They have been passed on from family member to family member, down generational lines. Ruth clung to

---

*   Not their real names.

her role in Sally's rescue for the rest of her life, and brought it up again and again to her children. She wanted them to believe in her as a heroine. She wanted them to know she was capable of a decent act. Some of her sons and daughters never reconciled Ruth's contradictions. Rachel, however, finally came to believe that her mother "did the best she knew how," no matter how many grievous mistakes she made as a mother and a human being.

SALLY HORNER'S FAMILY had to grapple with her sudden loss for the rest of their lives. Dolores Haze's husband, Dick Schiller, had to raise their child without her. But another woman had to reckon with the collateral damage of a father's abuse. That woman was Frank La Salle's daughter, known as Madeline.

Her mother, Dorothy, spent the Second World War working at the Brooklyn Navy Yard and rebuilding her life while her former husband was in prison. The choices she made as a newly divorced wife and single mother bear some resemblance to what was in store for Dolores Haze as Dick Schiller's wife. Madeline stayed with her mother during the summer months, and spent winters at her grandfather's house in Merchantville. After the war, when Madeline was ten, Dorothy met and married an army veteran several years her senior. He adopted Madeline and he and Dorothy had another child. Their marriage lasted until his death in 1986, almost four decades.

When her children were grown, Dorothy got a job with a small advertising firm, and then with Campbell's Soup, whose headquarters are still in Camden. She worked for the company for thirty years, retiring in 1991. Dorothy was also active in her local Baptist church for more than half a century, serving several years on the Board of Deaconesses.

When Dorothy died at ninety-two in 2011, her survivors included her children and almost a dozen grandchildren and great-grandchildren. The longer Dorothy lived, the more distance she put between her more settled, family-oriented existence and her turbulent early life with Frank La Salle. Madeline did not learn any details of her father's imprisonment until she was in her early twenties, newly married with children of her own. "There was an article in the newspaper, and my mother felt she had to tell me," she said in 2014. Knowing that her father was in prison did not repel Madeline. It made her curious. "I wanted to see him. I wanted to talk to him."

She reestablished relations with La Salle in the final year of his life, visiting him in Trenton State Prison along with her children, preschoolers at the time. He made model boats for the kids and leather pocket books for Madeline and her husband. When he was up for parole, sick with lung and heart problems, Madeline volunteered to have him live at her house should he be released early. That did not happen.

"When I looked at him, I could see a lot of myself in his face," Madeline said. "My husband picked it up right away." For those last months, Madeline did not clutter her relationship with questions of what La Salle had done to land him in prison. "We talked as father and daughter would talk," Madeline told me. "There wasn't a strain. He was just Dad. Truth be told, I never thought about whether he was guilty or not guilty."

Just as John Ray, Jr., became the conduit for Humbert Humbert's so-called confession, Madeline, unwittingly, became the keeper of Frank La Salle's version of the story. When I mentioned the word "abduction" to Madeline, she interrupted me with some force. "That's not the way he described it to me," she said. She then proceeded to parrot the version La Salle had presented in his appeals—a version the court had soundly rejected as fantasy.

FRANK LA SALLE never saw the outside world again. He appealed his sentence one final time, in 1962, and was again denied. He died of arteriosclerosis in Trenton State Prison on March 22, 1966, sixteen years into his sentence. He was, according to his death certificate, two months shy of turning seventy. The certificate listed him as "Frank La Salle III," the first time he ever used this sobriquet. That he died with his age shrouded in mystery and under a false name fits with the entire life of a man determined to conceal terrible truths.

TWENTY-FIVE

"Gee, Ed, That Was Bad Luck"

Two weeks after Sally Horner's death, on September 2, 1952, another sensational crime reported by the Associated Press caught Vladimir Nabokov's attention, and he filled another of his note cards. Unlike Sally's story, which merited a single parenthetical in *Lolita* but was seeded throughout the novel, this case got an entire paragraph at the beginning of chapter thirty-three. Humbert Humbert has returned to Ramsdale. Before making himself known in his former haunt, he stops off at the local cemetery, where he wanders as he ruminates upon his past. During his peregrinations, he stumbles across a particular sight:

> On some of the graves there were pale, transparent little national flags slumped in the windless air under the evergreens. Gee, Ed, that was bad luck—referring

to G. Edward Grammar, a thirty-five-year old New York office manager who had just been arrayed on a charge of murdering his thirty-three-year-old wife, Dorothy. Bidding for the perfect crime, Ed had bludgeoned his wife and put her into a car. The case came to light when two county policemen on patrol saw Mrs. Grammar's new big blue Chrysler, an anniversary present from her husband, speeding crazily down a hill, just inside their jurisdiction (God bless our good cops!). The car sideswiped a pole, ran up an embankment covered with beard grass, wild strawberry and cinquefoil, and overturned. The wheels were still gently spinning in the mellow sunlight when the officers removed Mrs. G's body. It appeared to be a routine highway accident at first. Alas, the woman's battered body did not match up with only minor damage suffered by the car. I did better.

Nabokov cleverly phrased it so that the reader isn't clear if Humbert actually stumbles across the murderer's grave or if he is merely thinking about the case as he looks at the graves. It has to be the latter, because Ramsdale is supposed to be somewhere in New England, an area Nabokov knew very well. The G. Edward Grammer case happened in Baltimore, a city Nabokov did not know at all. Nabokov's misspelling of Grammer's last name was deliberate, an opportunity for the noted literary prankster to sneak in another joke. It was also a sly reference to Humbert's professed intention, earlier in *Lolita*, to teach French grammar to Ramsdale's local children.

The text of Nabokov's surviving note card about the G. Edward Grammer case is close to, but not exactly, the final version. It includes the phrases "Gee Ed, that was bad luck" as well as "god bless our good cops!" But another wry aside about "Mrs. Grammar's new automobile" and Grammer's murderous actions did not make it into the final text: "ought to have doctored it first, Ed!"

One could see how this story, in tandem with Sally Horner's death, served as important inspiration for Nabokov. The Grammer case was a media sensation, capturing public attention for being an almost perfect murder. Grammer very nearly got away with it—except the Baltimore police noticed some details that did not add up, like a pebble jammed underneath the accelerator pedal.

The crime unfolded much as Nabokov described in *Lolita*. On the evening of August 19, 1952, Ed Grammer was getting ready to go back to New York City after a weekend with his wife and both of their daughters. Dorothy and the kids had moved to Parkville, a suburb of Baltimore, to care for her bereaved mother, while Ed remained in their Bronx apartment. The Sunday night routine was for Dorothy to drive Ed in their big blue Chrysler to Baltimore Penn Station, where he would give his wife some money for the week and catch the 11:28 P.M. train. For a few days after the "accident," Grammer insisted that Dorothy had dropped him off as usual, and that the last he'd seen her alive was at the train station.

But the facts didn't add up. The witnesses who saw the Chrysler speed down the hill along Taylor Avenue and sideswipe a telephone pole turned out to be patrolmen. For the victim of a car accident, Dorothy was astonishingly little-bruised in the areas they expected to be bruised, whereas her head had clearly been bashed in. There was blood in the driver's seat but the spatter wasn't substantial enough to suggest she had been killed on impact. More curious: Dorothy's purse and glasses were missing. When the pebble was discovered, pushing the accelerator forward, what seemed an accident transformed into murder, confirmed when Grammer confessed, at last, to Baltimore County police.

The fishbowl atmosphere intensified when reporters sniffed out the prospect of a mistress, which provided a motive for Dorothy Grammer's murder. But when they found her, she turned out to be a United Nations communications officer named Matilda Mizibrocky who swore she didn't know her beau was married, and they didn't print her name right away. Even the court hid her under the pseudonym of "Mary Matthews" so that she wouldn't be hounded further, and her testimony possibly tainted. It didn't work. Grammer's defense team was livid that the court tried to shield Mizibrocky from them, too, and hinder their ability to prepare their case.

It isn't clear if Nabokov followed the news after Grammer's arrest. The trial showcased further lurid details, and Grammer's execution by hanging in 1954 became an added spectacle because it was initially botched. But the main affair—husband murders wife, passes it off as car accident—was enough inspiration for him. The Grammer case clearly echoed the untimely death of Charlotte Haze, struck by a car after running away from the argument with Humbert where she learns of his true designs on her daughter.

The final line of the Grammer paragraph in *Lolita* reads with further chilling force. Grammer could not conceal his crime from the world after all. Humbert Humbert, systematically raping Dolores Haze for nearly two years on a cross-country odyssey, could, and did. No wonder he concluded: "I did better."

I bring up the Grammer case because it is another concrete example of Vladimir Nabokov drawing upon real-life crimes to help him with his novel. As with Sally Horner's kidnapping, the note card's survival indicates that Nabokov attached enough importance to the case that he wished people to know he did at some future point.

But the case also demonstrates Nabokov's extended interest in crime stories. This, too, he sought to deny in public; he was also openly critical of mystery novels despite his boyhood love of Edgar Allan Poe and Arthur Conan Doyle's Sherlock Holmes tales, and he called out Dostoevsky as a hack, though he taught *Crime and Punishment* to his Cornell students. He disdained those who would reduce *Lolita* to genre, yet a great deal of Nabokov's fiction relies on the tropes of crime and suspense: *Invitation to a Beheading* centers around a man waiting to be executed; *Despair* hinges upon a man ready to murder his double; and *Lolita,* of course, is about kidnapping and rape, and culminates in murder.

Which is also why Nabokov's interest, just over a month after *Lolita*'s American publication, in a third crime jumped out at me. As Véra told their close friend Morris Bishop when he telephoned with congratulations on the novel's success, in the week of September 12, 1958, Vladimir had become obsessed with reading up on the stabbing murders of Dr. Melvin Nimer and his wife, Louise Jean, in their Staten Island home. What fascinated Nabokov was that police initially treated their eight-year-old son, Melvin Jr., as a suspect. Even though strips of cloth found on the boy's bed suggested he had been restrained while his parents were murdered, Melvin's "unnaturally calm demeanor" raised red flags in investigators' minds, as did an apparent confession elicited during a mental health evaluation, and the lack of forced entry into the Nimer home.

But the presumed case against the little boy soon fell apart. No physical evidence linked Melvin to his parents' murders. And police learned that Dr. Nimer had left a set of spare keys at the hospital where he worked, which had vanished—thus answering the "lack of forced entry" question. The case remains unsolved to this day, but there were police detectives still claiming as recently as 2007 that Melvin Nimer was the best suspect in the case.

THE NABOKOVS WOULD VENTURE WEST one more time before Vladimir finished the *Lolita* manuscript. After so many years of work—five or six, depending on who was counting and who was listening—*Lolita* was nearly done, despite not being anywhere close to publication. This road trip also proved to be the longest Vladimir and Véra stayed away from the East Coast.

They left Ithaca in that still-reliable Oldsmobile in early April 1953. From there they headed toward Birmingham, Alabama, a pit stop en route to the Chiricahua Mountains in Arizona, where butterflies were supposed to be plentiful. What Nabokov discovered upon arrival in May was that the weather was too cold, the wind gusts too strong, for decent butterfly-catching. By the end of the month he and Véra had moved farther west, passing by several California lakes and ending up in Ashland, Oregon.

There the couple stayed from the first of June through the end of August, living at 163 Mead Street. When there were no butterflies to catalog, Nabokov was on a mad sprint to finish *Lolita,* burning his handwritten pages as soon as Véra typed them up. When their Oregon summer idyll ended, the Nabokovs wended their way back east via Jenny Lake and the Grand Tetons.

Once more, they were back in Ithaca at the start of September, and this time, the end of *Lolita* was in sight.

TWENTY-SIX

Writing and Publishing Lolita

On December 6, 1953, Vladimir Nabokov wrote a note in his diary, at the bottom of a page largely filled with numerical grades for the final assignment of his literature class. "Finished *Lolita* which was begun exactly 5 years ago." It was a finish line he'd spent many of those years never expecting to reach.

There were classes to teach at Cornell to pay the bills and to fund his summer trips to hunt butterflies. Other projects had also interrupted Nabokov's progress on *Lolita,* from translation work (*The Song of Igor's Campaign*) to the first version of his autobiography, which was published in 1951. *Lolita* ought to have been "a novel I would be able to finish in a year if I could completely concentrate upon it." Instead it emerged piecemeal, with him writing on index cards in the passenger seat of a car or lying in bed at night.

Nabokov had spent the summer of 1953 trip writing steadily, almost maniacally, dictating his prose to Véra, then "crumpling each old manuscript sheet once it had served its turn and discarding the pages out the car window or into a hotel fireplace." Nabokov put in sixteen-hour writing days over the course of the fall of 1953—on Cornell's dime—delegating the teaching and exam-marking to Véra.

He grew anxious about the manuscript as the pages piled up. In a September 29, 1953, letter to Katharine White at the *New Yorker,* Nabokov called the book an "enormous, mysterious, heartbreaking novel that, after five years of monstrous misgivings and diabolical labors, I have more or less completed." He was certain the *New Yorker* wouldn't want to publish an excerpt, but the magazine had a first-look agreement on anything Nabokov wrote, and he always listened to White's feedback, whatever the outcome. She liked it, but confirmed Nabokov's suspicions that the magazine wasn't the right home for an excerpt.

Now, the book was finished. An odyssey that did not, in fact, begin on December 6, 1948, but at least a decade earlier, as *Volshebnik,* or in 1947, when Nabokov wrote to Edmund Wilson: "I am writing . . . a short novel about a man who liked little girls—and it's going to be called *The Kingdom by the Sea.*" Nabokov knew he was writing a novel that could cause outrage and controversy. No wonder he attempted to destroy the manuscript at least twice that we know of.

The first time was in the fall of 1948. As Stacy Schiff recounted in her biography of Véra, Nabokov carried his manuscript to the trash cans behind his house on Seneca Street in Ithaca. When Véra realized what Vladimir was set upon doing, she raced out to stop him. Just before she got there, one of Nabokov's students at Cornell, Dick Keegan, chanced upon the scene. He saw Nabokov beginning to feed his manuscript sheets into a fire set near the trash cans. "Appalled, [Véra] fished the few sheets she could from the flames. Her husband began to protest. 'Get away from there!' Véra commanded, an order Vladimir obeyed as she stomped on the pages she had retrieved. 'We are keeping this,' she announced."

On at least one other occasion, when Nabokov wished to destroy the *Lolita* manuscript, Véra also stepped in as savior. It may well be that Nabokov's attempts to get rid of *Lolita* were more about performance than intent. As Robert Roper pointed out, "Véra came to the rescue because she was nearby; he did not start fires when his wife was out of the house." These acts made Véra a veritable Saint Joan[*] figure with respect to *Lolita*, sacrificing herself—if risking her husband's ire was a sacrifice—to step in and save what would be one of the most important works of literature in the twentieth century.

Nabokov later told the *Paris Review* of yet another instance of near-destruction, "one day in 1950." Once more, Véra "was responsible for stopping me and urging delay and second thoughts as, beset with technical difficulties and doubts, I was carrying the first chapters of *Lolita* to the garden incinerator." He may have mixed up the dates and this was the incident that Dick Keegan witnessed. Or there might have been an unrecorded instance where Véra saved the day.

Lolita was ready to be submitted to publishers, but there was a catch: Nabokov refused to put his own name to the novel. He asked Katharine White, in the same letter in which he solicited her feedback, whether book publishers would go

---

[*] Nabokov settled upon the "Lolita" sobriquet for his heroine very late in the writing process. Before then her name was "Juanita Dark"—a sly, Spanishized reworking of Jeanne d'Arc, or Saint Joan.

along with his request. She replied that "from her experience, an author's identity sooner or later leaked out." Still, Nabokov wanted to keep his identity secret, for the same reasons that spurred him to burn the manuscript pages of *Lolita* as he finished them. He believed being publicly associated with such an incendiary book might imperil both his literary and his teaching careers.

As the manuscript for *Lolita* made its way around New York publishing houses, Nabokov continued to insist that he publish under a pseudonym. His stubborn desire for anonymity may be one of the reasons why that first round of publishers decided against acquiring *Lolita*.

Nabokov's editor at Viking, Pascal Covici, rejected the manuscript. So, too, did James Laughlin, the New Directions publisher whom Nabokov worked with on *The Real Life of Sebastian Knight*, *Laughter in the Dark*, and *Nikolai Gogol*. Farrar, Straus and Simon & Schuster also came back with the same verdict: they didn't believe they could publish because it would be too expensive to defend in court on possible obscenity charges. Jason Epstein at Doubleday *did* want to publish, but the company president, once he got wind of what *Lolita* was about, overruled him.

The manuscript also detoured away from publishing offices, making its way into the literary world. The critic Edmund Wilson read half and expressed complicated feelings about the book in a letter to Nabokov ("I like it less than anything else of yours I have read"), perhaps because it reminded him too much of his own censorship battles after the publication of his novel *Memoirs of Hecate County*, which was banned and then pulped. Wilson's former wife, the novelist and literary critic Mary McCarthy, grew "negative and perplexed" by *Lolita*, but Wilson's present wife, Elena, liked the book. Dorothy Parker almost certainly read it, too, if a parody piece in the *New Yorker* featuring a character named "Lolita" is enough to go by.

Influential literary readers were all well and good, but their verdicts did not matter if the book never found a publisher. The last of the rejections by American book firms arrived in February 1955. To publish *Lolita*, Nabokov had to look outside of America, and beyond highbrow intellectual circles. Nabokov joked to Edmund Wilson several weeks later: "I suppose it will be finally published by some shady firm with a Viennese-Dream name." The joke became truth before the summer of 1955 was over.

MAURICE GIRODIAS was the founder and publisher of Olympia Press, best known for publishing books others wouldn't touch. Most of those books were, in fact, smut—badly written, hastily produced. Others got that label affixed to them, like Henry Miller's *Tropic of Capricorn* and *Tropic of Cancer*, J. P. Donleavy's *The Ginger Man*, and the pseudonymous *The Story of O* (revealed, decades later, to be the work of Anne Desclos).

Nabokov's Europe-based agent, Doussia Ergaz, submitted *Lolita* to Girodias because of his work as an art-book publisher. She didn't seem to know much about the seedier side of Olympia Press. Girodias was fully aware of the literary value of Nabokov's work, and what a boon it would be for Olympia's list. Girodias offered on the book in mid-May 1955. Ergaz then wrote Nabokov: "He finds the book not only admirable from the literary point of view, but he thinks that it might lead to a change in social attitudes toward the kind of love described in *Lolita*, provided of course that it has this authenticity, this burning and irrepressible ardor."

Nabokov went along with Girodias's misapprehension

about there being a social aim to the novel because he was relieved *Lolita* had at last found a publisher. That relief dissipated quickly, once he realized the contract he'd signed on June 6, 1955, better resembled a devil's bargain. Nabokov's new publisher had mistaken the author for his creation, thinking Nabokov drew upon some perverse experience. Girodias also insisted the novel be published under Nabokov's name, and Nabokov did not feel he had the leverage to object, when the alternate option was no publication at all. Nabokov also did not see galley proofs until it was too late to make changes, which vexed a man known for his fastidiousness to no end. Olympia Press published *Lolita* on September 16, 1955, but Nabokov did not discover that it was out for several weeks. And the published version, as Nabokov feared, was riddled with errors.

What most infuriated Nabokov, however, was Girodias's blithe attitude about copyright and about paying him what he was owed. The publisher had registered *Lolita*'s copyright in Nabokov's name as well as to Olympia Press. Nabokov did not discover the joint copyright registration until early 1956, and because of American copyright laws at the time, he had just five years to republish *Lolita* in America or else the novel would fall into the public domain.

The Copyright Office in Washington advised Nabokov to get a "quit-claim"—a formal renunciation of copyright. The publisher did not reply at first, then dragged his feet throughout 1956 and 1957. As Nabokov later recalled, "From the very start I was confronted with the peculiar aura surrounding [Girodias's] business transactions with me, an aura of negligence, evasiveness, procrastination, and falsity."

Girodias also had a pesky habit of failing to pay royalties or to send statements. Thus, Nabokov saw no money from *Lolita* over the first two years of publication, despite strong sales in

France. In October 1957, he had finally had enough of Girodi-as's prevarications and shady dealings, telling him the deal was off and that as a result, all rights reverted back to him. Giro-dias paid what was owed (44,220 *"anciens* francs"), and Nabo-kov let the matter go. Girodias, however, soon reverted back to his nonpayment ways, and Nabokov's irritation increased. He needed the money, but above all, he needed to be free of Olym-pia in order to publish *Lolita* the way he had always wished.

Fortunately for Nabokov, his mood was about to lighten. *Lolita* was about to find, at long last, a home in America.

ON AUGUST 30, 1957, Nabokov received a letter from Walter Minton, president and publisher of G. P. Putnam's Sons. "Be-ing a rather backward example of that rather backward spe-cies, the American publisher, it was only recently that I began to hear about a book called *Lolita,*" Minton wrote. After some more preamble, he got to the point: "I am wondering if the book is available for publication."

Minton, in his early thirties, had succeeded his father, Melville, as publisher two years earlier, and within months established his taste for novels too controversial for other pub-lishers. Putnam published Norman Mailer's second novel, *The Deer Park,* which had been turned down by his option pub-lisher, and several others, for a graphic description of oral sex that each publisher feared would run afoul of obscenity laws. This passage did not deter Minton, who authorized Putnam to run newspaper ads declaring *The Deer Park* was "The Book Six Publishers Refused to Bring You!"

Minton enjoyed being part of the cultural conversation, es-pecially when there was a chance that the conversation would offend people. In hindsight, it made sense he ended up publishing *Lolita*. The delicious thing is that he learned of the novel from an unlikely source: his then-mistress Rosemary Ridgewell, a showgirl at a Midtown Manhattan nightclub called the Latin Quarter. Ridgewell had read excerpts of *Lolita* in the *Anchor Review*. "I thought Nabokov had a very interesting way of writing, very, you know—crystalline?" said Ridgewell in 1958.

Minton, in turn, discovered the pages at her Upper East Side apartment. "I woke in the middle of the night and there was this story on the table. I started reading," he recalled in early 2018, sixty years later. "By morning, I knew I had to publish it." (Ridgewell was in line for a tidy payday for her literary scouting efforts, per a standing Putnam policy: the equivalent of 10 percent of an author's royalties for the first year, plus 10 percent of the publisher's share of subsidiary rights for two years.)

When Nabokov received Minton's letter, he had all but given up on *Lolita*'s publishing prospects in America. For more than three years, multiple publishers had expressed interest, only to back off. Now it irked him that Girodias might be in line for a significant payout when he had been so slow with the initial *Lolita* royalties, and then did not bother to pay further monies Nabokov was owed. In the two years since *Lolita* first appeared in book form, he was desperate to reap the financial rewards—as well as to get the critical attention he deserved.

Lolita had been banned in France, excerpted in the *Anchor Review*, praised by the novelist Graham Greene, excoriated by the literary editor and critic John Gordon, and bought in droves by those willing to smuggle copies back into America. All manner of people benefited from *Lolita*, whether to praise or denounce it, but Vladimir Nabokov had hardly earned a dime for his years of creative labor.

Minton's letter augured a change in fortune. Nabokov

wrote back on September 7 to say Minton was free to negotiate with Olympia Press, though "I would have to give my approval to the final arrangements." Nabokov added a warning: "Mr. Girodias, the owner of Olympia, is a rather difficult person. I shall be delighted if you come to terms with him." Minton did not seem fazed by Girodias's unsavory reputation. Nor was the Putnam publisher perturbed by the prospect of defending *Lolita* all the way to the Supreme Court, if necessary, though he cautioned Nabokov that he could not make such a "blanket guarantee"—rather, he wrote that it was more prudent to "present the book in such a way as to minimize its chance of prosecution."

Minton also wondered whether *Lolita* could, in fact, fall into the public domain. He said as much when he met the author in Ithaca, braving a snowstorm to do so. Nabokov told Minton that he knew "at least three or four thousand copies" of the Olympia Press edition of *Lolita* had been sold in the United States. Minton explained, "I said to him, 'Don't ever open your mouth about that to anybody because if it ever became established your copyright wouldn't be worth *beans*."

Nabokov kept his mouth shut about these extra sales. He took more time to swallow his pride with respect to Girodias. Much as he wanted to be financially free of his former publisher—going so far as to declare the original contract null and void in light of Girodias's inability to abide by the terms—he agreed, grudgingly, with Minton that it was better to allow Girodias a stake in *Lolita*'s American publication so they could be sure to get the book out as soon as possible.

As Minton explained to him over the winter of 1958, publishing *Lolita* when interest was high would make it more likely that the courts would rule in Nabokov's favor: the many articles, vociferous discussion, and chatter would demonstrate this was a book of high literary merit, not low smut. Should Nabokov delay in publishing *Lolita* in America to resolve his dispute with Girodias, the favorable publicity could evaporate—and so would the potential for a great financial windfall, whatever a court of law might decide.

Nabokov saw Minton's logic. He wrote the publisher back in early February 1958 to agree to terms (including the 50/50 royalty split with Girodias, with each receiving 7.5 percent of the hardcover proceeds). Minton cabled Girodias on February 11 and received Nabokov's signed contract on March 1.

By the time of the novel's American publication date on August 18, 1958, it was clear to all, but most especially to Nabokov, that he was about to be vaulted from literary obscurity, and that *Lolita* was about to arrive with hurricane-level force.

VLADIMIR AND VÉRA NABOKOV left Ithaca on another road trip in the summer of 1958. Whether to fend off nerves or steel themselves for what was to come, the couple traveled more than eight thousand miles in search of butterflies. Nabokov had also decided to take a leave of absence from Cornell beginning in the fall because of all of the prepublication demands for *Lolita*. They returned to New York in early August, in time for a press reception at the Harvard Club. Véra recorded her impressions of the evening, and of her husband, in their shared Page-a-Day diary: "Vladimir was a tremendous success . . . amusing, brilliant, and—thank God—did not say what he thinks of some famous contemporaries."

On publication day, Minton sent the following telegram to the Nabokovs: "EVERYBODY TALKING OF LOLITA ON PUBLICATION DAY YESTERDAYS REVIEWS MAGNIFICENT AND NEW YORK TIMES BLAST THIS MORNING PROVIDED NECESSARY FUEL TO FLAME 300 REORDERS THIS

MORNING AND BOOK STORES REPORT EXCELLENT DEMAND CONGRATU-
LATIONS."

Minton was referring to Elizabeth Janeway's rave review, which ran on Sunday, August 17, in the *New York Times Book Review*. She described the novel as "one of the funniest and one of the saddest books of the year" and declared that it was anything but pornographic: "I can think of few volumes more likely to quench the flames of lust than this exact and immediate description of its consequences." Janeway's positive reaction, plus the increased demand, would compensate for Orville Prescott's pan in the daily paper on publication day proper, August 18.

The reorder number from retailers zoomed up to 6,777 in the first four days after its publication. By the end of September, *Lolita* was atop the *New York Times* bestseller list, having sold more than eighty thousand copies. It remained at number one for the next seven weeks. Six months after ensuring *Lolita* would be published in America without legal hassle or copyright consequence, Nabokov and Minton's mutual investment was paying clear and major dividends. There were more riches in store for the Nabokovs. On Minton's recommendation, they retained Irving "Swifty" Lazar to sell film rights to *Lolita* to Stanley Kubrick for $150,000.

Véra was the one who kept track of it all, recording every bit of *Lolita*-related news in the months immediately preceding and following the novel's publication in the Page-a-Day diary. Nabokov, on the other hand, seemed "supremely indifferent—occupied with a new story" and with cataloging his summertime butterfly-hunting bounty. Or at least that was the guise he adopted, as described by his wife. As the deluge of letters, interview requests, and subsidiary rights inquiries streamed in, Nabokov wrote his sister: "[All this] ought to have happened thirty years ago. . . . I don't think I shall need to teach any more."

Nabokov proved correct. The indefinite leave of 1958 became permanent retirement from Cornell at the end of 1959, when he was sixty. *Lolita*'s success enabled his permanent break with the United States, though he would continue to insist, years after moving to the Montreux Palace hotel in Switzerland, that he might return. But Switzerland's advantageous tax laws were too much of a boon—as was a greater sense of control and privacy, as "Hurricane *Lolita*" grew stronger and louder. Now that Nabokov could afford to write full-time, thanks to his most American book, he could embark upon his next phase: as a literary celebrity in voluntary exile, as opposed to the peripatetic refugee. He would see the country that gave him shelter, sanctuary, and the source material for his most famous novel again only a handful of times.

Lolita moved far beyond the bestseller list to become a cultural and global phenomenon. The template was in place for generations of readers to be taken in by Humbert Humbert, forgetting that Dolores Haze was his victim, not his seducer.

At the time, no one noticed that *Lolita* was published in the United States on the sixth anniversary, to the day, of Sally's death. And no one made the connection between fictional nymphet and real girl for several more years.

TWENTY-SEVEN

Connecting Sally Horner
to Lolita

Peter Welding was a young freelance reporter in 1963, still a few years shy of thirty. A Philadelphia native, he had already built up a solid series of bylines, mostly for music magazines like *Downbeat*. He was also a fledgling music producer, founding Testament Records that same year to issue old and new jazz, gospel, and blues recordings. Welding had moved to Chicago to further his producing career, but not before setting out to tell a story that unfolded across the river from his hometown.

Welding was born in 1935, two years earlier than Sally Horner; no doubt her kidnapping and rescue made a large impression on him as a teenager. Welding remembered reading of Sally's plight in his local newspapers, the *Philadelphia Inquirer* and the *Evening Bulletin*, and decided to examine the

LOLITA
HAS
A
SECRET,
SHHH!

ARTICLE BY
PETER J. WELDING

THE TRUE SCOOP
BEHIND MR. NABOKOV'S
FAMOUS NOVEL

Image accompanying Peter Welding's November 1963 article for Nugget.

parallels between Sally's story and *Lolita*. He zeroed in on the same parenthetical phrase that, decades later, first caught the attention of Nabokov scholar Alexander Dolinin and, then, me. He compared specific events that occurred in *Lolita* to what happened to Sally. The results were published in an unusual venue: the men's magazine *Nugget*, racier than *Esquire* or *GQ* but more prudent than *Playboy*.

Nugget had a knack for publishing stories with literary connections. It helped that its editor at the time was Seymour Krim, who was affiliated with the Beat Generation, hanging around with Jack Kerouac, Allen Ginsberg, and Neal Cassady, though unlike them, he never wrote poetry or fiction. Krim also embraced the New Journalism ethos while working at the *New York Herald Tribune* with future stars Tom Wolfe, Jimmy Breslin, and Dick Schaap. *Nugget*, under Krim's editorial guidance, featured stories and articles by Norman Mailer, James Baldwin, Otto Preminger, William Saroyan, Chester Himes, and Paddy Chayefsky—and that was just in 1963.

Despite shooting for high literary quality, *Nugget* wanted to reach a mass audience—though Krim and his staffers sabotaged their own efforts by failing to stick to a regular publishing schedule. Just five issues of *Nugget* appeared in 1963;

Nugget's frequency during Krim's editorial tenure might be best described as "bimonthly-ish." Welding's story, "*Lolita* Has a Secret—Shhh!" ran in the November issue.

The piece opened with a summary of both *Lolita* and Sally's abduction, which Welding had gleaned entirely from news reports in his hometown papers, the *Inquirer* and *Evening Bulletin*. He recounted the five-and-dime meet-up, La Salle's threat of juvenile incarceration if Sally didn't go along with his plan, the basic outlines of the twenty-one-month-long cross-country trip, Ruth Janisch's role as rescuer, and Sally's eventual recovery.*

After Welding finished with his summary, he declared that the Horner case and *Lolita* "parallel each other much too closely to be coincidental." Welding took particular note of Sally's fear of being sent to reform school, which he compared to Humbert's declaration, roughly halfway into the novel, that "the reformatory threat is the one I recall with the deepest moan of shame." And further:

"... What happens if you complain to the police of my having kidnapped and raped you? ... So I go to jail. Okay. I go to jail. But what happens to you, my orphan? ... While I stand gripping the bars, you happy neglected child, will be given a choice of various dwelling places, all more or less the same, the correctional school, the reformatory, the juvenile detention home. ..."

Welding also drew a comparison between Humbert marrying Charlotte Haze to gain access to Dolores and La Salle claiming

---

* Oddly, Welding referred to the girl as "Florence 'Sally' Ann Horner." It is a mystery why he gave her a middle name that was never reported and did not exist. The error continued to propagate, also cited by Alfred Appel, Jr., in *The Annotated Lolita*.

to be married to Sally's mother. Welding further speculated that the actions of Humbert's housekeeper at Beardsley, Mrs. Holigan, might be based on the actions of Ruth Janisch ("on the rare occasions where Holigan's presence happened to coincide with Lo's, simple Lo might succumb to buxom sympathy in the course of a cozy kitchen chat").

Finally, Welding arrived at his own smoking gun: that neon-light parenthetical late in *Lolita:* "in this single reference [Nabokov] has all the essentials—the full names, the ages of the pair, La Salle's occupation, the date of the abduction—neatly packaged in one sentence," which suggests thorough familiarity rather than casual knowledge.

Eventually Welding seemed to tire of the compare/contrast—"More parallels could easily be shown, but to what further purpose?"—but he felt confident enough to conclude: "The plot line of *Lolita* derives in large measure from the case and is solidly based on actual fact. . . . The conclusion is almost inescapable: Nabokov has inserted the reference to the LaSalle-Horner story either as a conscious (or unconscious) acknowledgment of a primary source material, or as a shrewd maneuver to provide himself legal protection."

I'm not certain what Welding meant here. He didn't elaborate further in the piece, so I can't know for sure, but perhaps he believed, based on this statement, that lawyers, already made nervous by *Lolita*'s worldwide controversy, legal challenges, and publication bans, insisted Nabokov insert the reference to avoid an additional legal issue, such as a plagiarism charge.

Welding made several mistakes in his piece. First, he got Humbert's parenthetical comment wrong, suggesting that it was from the vantage point of Mrs. Chatfield, the mother of one of Lolita's classmates whom Humbert meets in a hotel lobby upon his return, after more than five years away, to Ramsdale. Still, Welding understood the importance of the Sally Horner aside decades before anyone else.

More baffling to me was his omission of Sally Horner's fate from the article. Was Welding unaware that Sally died in a car accident a decade earlier? Could the "shrewd maneuver" line have referred to possible grounds for Sally to sue, had she been alive? In omitting Sally's death, he missed an opportunity to compare her fate to Charlotte Haze's death by car accident. What was beyond Welding's grasp, though, was the direct proof Nabokov knew all about Sally Horner, since the transcribed and reworked wire report that is part of his archives at the Library of Congress was not available to the public in 1963.

Lastly, Welding theorized, with the careful, awkwardly worded hedge, that "it is not unlikely to suppose" that Nabokov did not begin to work on *Lolita* "until 1950, under the stimulus of the stories of the LaSalle-Horner case. All evidence, in fact, would seem to support this position." This is false, since Nabokov's own December 1953 diary entry, celebrating the completion of the manuscript after five years of work, refutes Welding. We also know this supposition is false thanks to the existence of *The Enchanter*.

However, 1950 was around the time Nabokov came close to junking what was still then called *The Kingdom by the Sea*. Welding was off base but his suppositions make sense: whenever Nabokov first learned of what happened to Sally Horner, that knowledge helped him to transform a partial manuscript primed for failure into the eventual, unlikely, staggering success of *Lolita*. If such knowledge was publicized, it would not look good for Nabokov—rightly or wrongly—to be seen as pilfering from a real girl's plight for his fictional masterpiece.

THE NOVEMBER 1963 ISSUE of *Nugget* came and went without much notice. It attracted nowhere near the attention of the release of the film version of *Lolita* a month later, or the millions of sales garnered by the novel to date. But one person did pay attention: a *New York Post* reporter named Alan Levin.

Levin eventually became an award-winning documentary filmmaker, working with his son, Marc, and with Bill Moyers, on films shown by PBS and HBO. But he cut his journalistic teeth for the Associated Press and then joined the *Post* in the late 1950s, where his reporting on organized crime garnered him a Pulitzer Prize nomination.

It's not clear whether Levin, thirty-seven at the time, found an advance copy of the November issue of *Nugget* on his own or if someone tipped him off. However he got ahold of Welding's story, Levin knew there was one of his own to write—could it be true that *Lolita* owed its plot to a sensational kidnapping? And if it was true, what would the great Vladimir Nabokov have to say on the matter?

Levin posted a letter to Nabokov on September 9, 1963, which arrived in Montreux, Switzerland, just four days later. The official Nabokov response, written and signed by Véra, reached Levin soon thereafter. Levin's story for the *Post*—"Nabokov Says 'Lolita' Is More Art Than Life"—ran on September 18, 1963, a day after Levin received Véra's letter.

The article began with a provocative lede: "Is it possible that Humbert Humbert was a 50-year-old Philadelphia auto mechanic and his nymphet, Lolita, an 11-year-old from Camden, NJ?" Levin quoted just enough of Véra's letter to make the Nabokov case plain, and enough of Welding's *Nugget* piece to answer the question. But it's helpful to read Véra's letter to Levin in full, as it offers a fascinating window into her (and

Nabokov's) thought process. Her response also carries an air of protesting too much:

> *Montreux, September 13, 1963*
> *Palace Hotel*
>
> *Dear Mr. Levin,*
>   *My husband asks me to thank you for your letter of September 9. He has not seen the article in* Nugget, *which makes it difficult for him to answer your letter. At the time he was writing LOLITA he studied a considerable number of case histories ("real" stories) many of which have more affinities with the LOLITA plot than the one mentioned by Mr. Welding. The latter is mentioned also in the book LOLITA. It did not inspire the book. My husband wonders what importance could possibly be attached to the existence in "real" life of "actual rape abductions" when explaining the existence of an "invented" book. He is particularly curious as regards the meaning of Mr. Welding's statement about "a shrewd maneuver to provide himself legal protection." Legal protection against what?*
>   *Had he read Mr. Welding's article, my husband might have been able to give you more pertinent comment although he fails to see what importance that article could possibly have.*
>   *Sincerely yours,*
>   *(Mrs. Vladimir Nabokov)*

Véra's letter showcases the many roles she played as "Mrs. Vladimir Nabokov": defender of her husband, curator of a singular line of vision about Nabokov's work that put his creative genius above everything else, and master obfuscator when presented with anything that dented the Nabokov myth. Véra, again, showed herself to be the consummate brand manager for Vladimir. Any speculation that *Lolita* could be inspired by a real-life case went against the single-minded Nabokovian belief that art supersedes influence, and so influence must be brushed off.

How the Nabokovs handled Levin's letter, and by extension Welding's article for *Nugget,* is a window into their maddening, contradictory behavior when anyone probed *Lolita*'s possible influences. They denied the importance of Sally Horner but acknowledged the parenthetical. They mentioned a "considerable number" of case histories, but only Sally's is described in the novel.

Véra's stubborn insistence that the Sally Horner story "did not inspire the book" is akin to trying to drown out a troublesome argument with the braying of one's own voice. Though it worked, since Levin did not push back—at least, not that we know of.

That Véra claimed the Nabokovs had not seen the *Nugget* piece is odd on several counts. First, the author's main archive at the New York Public Library contains over half a dozen boxes' worth of newspaper clippings about *Lolita,* starting from its original 1955 publication by Olympia Press through its American publication by Putnam in 1958, well into the 1960s and early 1970s. The Nabokovs subscribed to several clipping services based in New York and Paris. They seemed to have kept every review in every possible language they spoke or read—English, French, German, Italian, Russian—whether good or bad, critical or full of praise, defending its content or wishing the book banned from the earth.

There is also an entire box of clippings related to the *Lolita* film, beginning with who would be cast to play Dolores Haze. The Nabokovs even kept a copy of the August 1960 issue of *Cosmopolitan,* featuring Zsa Zsa Gabor, at the ripe old age of forty-three, dressed up as twelve-year-old Lolita in a straining baby-doll nightgown, an apple in her hands, licking her lips in equal parts faux-innocence and come-hither enticement. And other fashion and girlie mags from France and Italy, each with photo shoots of starlets garbed in Lolita-like frocks as a pictorial audition for a film part they desperately wanted to play.

It staggered me, this voluminous collection of *Lolita* ephemera. And yet, there was no sign of the relevant issue of *Nugget.* Compared to other periodicals collected and kept by the Nabokovs, *Nugget* was not so obscure. Its absence is telling because it is part of a larger absence in Nabokov's archives: any reference whatsoever to Sally Horner.

Véra Nabokov, in her letter to Al Levin, emphasized that Sally's abduction "did not inspire the book." Moreover, she insisted Nabokov "studied a considerable number of case histories . . . many of which have more affinities with the *Lolita* plot than the one mentioned by Mr. Welding." Even if that was true, the statement was disingenuous. For only two of the "considerable number of case histories" were explicitly mentioned in *Lolita:* the story of G. Edward Grammer, and the story of Sally Horner.

Nabokov must have had a reason to hold on to those two index cards and not burn them, as he had burned handwritten pages of the manuscript. He had been compelled to write notes on both cases, and in particular the death of Sally Horner. He included the parenthetical reference in the novel when he could have left out any mention altogether. Sally's story mattered to Nabokov because *Lolita* would not have been finished if he hadn't read of Sally's kidnapping.

The Nabokovs' behavior could, I suppose, be attributed as much to carelessness as willful obfuscation. Stacy Schiff, Véra's biographer, strongly advised against reading anything specific into Véra's blanket denial to Levin. Schiff told me that Véra's letter "reads like everything else [the Nabokovs said] about the primacy of art. It's a realm unto itself, and everything else is on some pedestrian or insignificant level." Véra, Schiff said, dismissed anything that could be perceived as a "mandarin influence on high art." The everyday needed to be discarded at the altar of creative imagination.

Except Vladimir and Véra were not careless people. His art, and her management and protection of his art, was all about command and control, about rejecting interpretations that did not fit with their vision. If art was to prevail—and for the Nabokovs, it always did—then explicitly revealing what lay behind the curtain of fiction in the form of a real-life case could shatter the illusion of total creative control.

Véra's denial by letter had to be definitive to make pesky tabloid reporters slink away without investigating the matter more deeply. Levin published his piece in the *Post*, but it was soon forgotten, setting the template for further neglect of the Sally Horner case. Andrew Field, in his 1967 critical biography *Nabokov: His Life in Art*, merely cited the parenthetical as "an actual case of a Philadelphia mechanic who took an eleven-year-old Camden girl to Atlantic City."

Alfred Appel, in his annotated *Lolita* published in 1970 (as well as in the revised 1991 version), dutifully footnoted the reference to her, but failed to mention Sally by name. Brian Boyd's definitive biography of Nabokov was better, noting "'a middle-aged morals offender' who abducted fifteen-year-old Sally Horner from New Jersey and kept her for twenty-one months as his 'cross-country slave'"—but misstated Sally's age at her abduction by four years.

Vladimir Nabokov's otherwise scrupulous archive of *Lolita*-related clippings failed to include anything about Sally Horner because if it had, then the dots would connect with more force, which would upset the carefully constructed myth of Nabokov, the sui generis artist, whose imagination and gifts were far superior to others'. It's as if he didn't trust *Lolita* to stand on its own against the real story of Sally Horner. As a result, Sally's plight was sanded over, all but forgotten.

TWENTY-EIGHT

"He Told Me Not to Tell"

Decades after Ruth Janisch gently coaxed Sally Horner to make the long-distance telephone call that freed her from Frank La Salle, Ruth was having tea with her daughter Rachel. After years of estrangement, Rachel had decided she wanted a closer relationship to her mother.

In Sally's story, Ruth was a heroine whose actions changed the course of a girl's life forever. But to her children, Ruth was a more complicated, infuriating, mercurial, manipulative creature, whose actions led to long estrangements. That troubling Ruth was not yet in full bloom in 1949 and 1950. Much of her aberrant behavior was still in the future. Rachel described her mother's life philosophy to me: Ruth would meet someone and say something along the lines of "Hello, my name is Ruth. What can you do for me?"

Years into adulthood, some of Ruth's children would make

peace with the woman she was. Yes, Ruth had done terrible things in the past. She had looked the other way when her children were abused, physically, emotionally, and sexually, by the men in her life, be they husbands or short-lived romantic partners. Ruth had, at times, enabled that abuse by not believing her children and choosing, instead, to believe the men. One of them ended up as Rachel's first and only husband.

Ruth was working at a bus station in the Bay Area at that time, around the early 1960s. One of her coworkers had two sons, whom Ruth decided must meet her daughters. Ruth wanted the older boy for herself, but she thought the younger one, still a teenager, would be perfect for Rachel. Instead, the younger boy expressed no interest, and the older one gravitated toward Rachel in a way that made her wonder, much later, if there was something more calculated at play.

Rachel grew certain that her mother had made some sordid arrangement with the older boy. That in order for him to have access to Rachel, he had to have some romantic involvement with her mother. Ruth also goaded her daughter about him at the time, saying she couldn't possibly land a man like him. Rachel would not grasp the impact of her mother's verbal abuse for years. Then, she thought Ruth's behavior was normal.

Rachel did "land" the boy, became pregnant, and she then married him in haste and moved away from home. Seventeen years, three children, numerous moves, and countless beatings, rapes, and threats to her life later, Rachel managed to break free. "It was less a marriage than extended captivity," she said. When she dared to speak up for herself, her husband punished her. He repeated the pattern she knew too well from childhood, when confessing that something hurt her caused more hurt, psychologically from Ruth and physically from her husbands or partners.

Once Rachel's divorce was final, in the late 1970s, she

found a job near where her mother lived. She also thought about what kind of relationship she wanted with Ruth. Because Rachel, despite the past, *liked* her mother. They shared a love of gardening and of books. As adult women, they could converse, if not as equals then at least on a similar plane. Rachel decided she could handle visiting Ruth at least once a week for tea. The visits were calm at first. She felt herself understanding her mother better. She felt she had enough emotional distance to appreciate Ruth, the woman, and leave the baggage of neglect and abuse behind.

What Rachel created in this new relationship with her mother was a cocoon where old wounds could be erased. But the cocoon turned out to be an illusion. That it broke apart, Rachel realized, should not have surprised her as it did.

One afternoon, Rachel turned up at Ruth's house and found her mother immersed in her scrapbooks. They were a living testament to Ruth's belief that she mattered. Subsequent visits by Rachel and other daughters led to them discovering that the scrapbooks contained items Ruth had no business possessing, items she had pilfered from her children without their immediate knowledge, but Rachel wasn't aware of that yet. As mother and daughter sat for tea, one scrapbook lay between them. The one Ruth had been working on earlier that morning.

Lying next to the scrapbook was an old paperback novel that Rachel didn't recognize.

Ruth pointed to a newspaper clipping in the scrapbook. "Do you remember this little girl?" Ruth moved her finger to the headline. "Do you remember that girl, Sally Horner?"

"I do remember," said Rachel.

"And do you remember Frank La Salle, the man who kidnapped her?"

"I do remember," Rachel said again.

Rachel told me that she felt herself grow cold. But Ruth did not notice, or even understand, her daughter's reaction.

"Wasn't that quite the story?" Ruth carried on. "They're mentioned in this novel, *Lolita*!"

Rachel hadn't heard this story before. She would hear a different version of it years later from her sister Vanessa, who had read the novel—that same tattered paperback—at Ruth's insistence. Vanessa, sixteen at the time, was puzzled at first at her mother's insistence she read the novel. Then Ruth explained: "*Lolita* tells the story of a girl named Sally Horner. A girl who died just before you were born. A girl I helped rescue from a man named Frank La Salle." This was no mere novel, but literary validation of Ruth's sense of self, that her single act of decency had larger heroic meaning. Vanessa couldn't help but read *Lolita* with her mother's words echoing in her head. But she also couldn't read it without thinking of the ways in which Ruth had wronged the family.

Staring at the scrapbook, Rachel knew she had to speak up now or she would never be able to say the words again.

"Mom, there's something I need to tell you. Something I never told you. Frank La Salle didn't only molest Sally. He also molested me."

ONE AFTERNOON while Sally Horner was at school, Frank La Salle invited Rachel over to his trailer. At age five, Rachel wore her white-blond hair in pigtails, a ribbon adorning the left plait. She put on her favorite striped black-and-white dress every chance she got. One portrait of Rachel from that time depicts her as the ideal of innocence. Her smile is trusting, guileless. Her expression is pliant, hinting at a gullible streak that would plague her repeatedly in adulthood.

She might have been by herself, or with one of her sisters. What she most recalled was that Frank was nice to her. He seemed to understand what Rachel wanted. That she saw what Sally had—toys, games, a father's love—and didn't have them. Not the toys. Not the games. And her father, George, was a remote presence, more comfortable with adults than with his children.

Of all the things Sally possessed, what interested Rachel most were her crayons and coloring books. They reminded Rachel of an earlier stay at a Minnesota hospital when she was three, quarantined for months with rheumatic fever. She went to school at the hospital because she wasn't allowed to go home. There were troubles, both at home and in the hospital, the kind so traumatizing Rachel would not remember them for decades without the trigger of unexpected smells and brief image flashes. But what came back to her now was the memory of her parents visiting with big smiles and good cheer, and of learning how to draw with crayons.

Frank made Rachel a deal: She could play with anything of Sally's. Anything at all. But she had to do favors for him first.

"Essentially, what he wanted from me was to give him a blow job," Rachel said. "So I did."

Rachel remembered a single incident. There may have been others that she blocked out. She didn't remember it until after she was married, and her husband wanted the same thing. The sexual act seemed disgusting to the newly married Rachel, and it would take her years to learn it flowed from normal, healthy desire.

She only realized the enormity of what Frank had done to her when she stumbled across a pamphlet at the local library, long after the end of her marriage. The pamphlet was about child molestation. Its title: "He Told Me Not to Tell."

Frank La Salle had told Rachel not to tell. But Rachel would have stayed silent regardless. The girl's earlier hospital experience taught her not to trust adults. They could abandon you for months. They could leave you alone and never tell you when you might come home. And when you did come home, you didn't know what awaited. Whether it was a safe haven or a recurring nightmare.

While telling the story to me nearly forty years after the fact, Rachel would recall that when she told her mother what La Salle had done, a barrier went up right away between herself and Ruth. Like an electrified fence where venturing too close might lead to a surprise shock. She felt her mother shut down. Rachel decided to change the subject to safer territory. She had taken a risk and it did not work. When that happened, as it had so many times in her past, the best thing to do was to be like a turtle. Retreat within the shell and never reveal your vulnerable self again.

Afterwards, she and Ruth would never be as close. (There would be other visits, including a reunion of all but one of Ruth's children in 1998, six years before Ruth died.) Rachel sensed her mother must have known, on some level, why there was a new breach between them. Perhaps, Rachel theorized, her mother had helped Sally Horner because she knew she could not save her own children from nearby monsters. But letting her thoughts run in this direction was too much for Rachel. She suspected Ruth never allowed herself to contemplate her complicity in the damage done to her own children.

No wonder Ruth acted as if the conversation never happened. The subject never came up between them again. Neither did *Lolita*.

TWENTY-NINE

Aftermaths

S ally Horner's premature death cast a pall on the lives of those who knew her best and loved her most. Ella, who did not display much of her inner turmoil while Sally lived, buried it completely after her daughter's death. She could not think of another way to handle it besides burying her emotions, but at least she did not have to bear her grief alone. The Panaros remained nearby. Ella could watch her granddaughter, Diana, grow up.

A year before Sally died, Ella had connected with a new partner: Arthur Burkett, a Camden native five years her junior, who moved into 944 Linden Street. They decamped for Pennsauken, five miles away, within a year of the car accident that killed Sally, and in the early 1960s migrated west to Palo Alto, California—less than an hour's drive away from the San Jose trailer park where Sally had been rescued. Burkett found work as a groundsman for a local college, and Ella and Ott, as he was called, made their union legal in January 1965.

Five years later, Burkett was dead. His tractor had overturned on him as he cut grass on a steep hill on the college grounds. While he was recuperating from the accident, doctors discovered he had gastric cancer, which had already spread to his liver. Without any other ties to Palo Alto, after he died Ella returned to New Jersey to be closer to her family.

When she visited her family, she never spoke of what happened to Sally. She didn't speak much of the past at all. There didn't seem to be much purpose in revisiting painful memories with a generation that hadn't been born when the worst had happened. Ella preferred to play cards—gin rummy was a favorite—or talk about what books she liked to read. Mystery novels, in particular.

Ella's silence about what happened to her younger daughter carried over to the next generation. Diana didn't learn the truth that her dead aunt Sally had been kidnapped until she was in her teens. She didn't remember the exact details of the conversation she had with her father about Sally, but she recalled it didn't last long. A mere recounting of the basic facts: that before her aunt died, she had been taken by a stranger posing as her father. The rest was left up to Diana's imagination.

Diana was only four years old when Sally died. Her parents were chiefly concerned with making sure their daughter had a happy childhood, and hiding their hurt, however deep the wounds ran. Just as Susan didn't speak much about her younger sister, neither did Al speak of his experiences in World War II. It seemed easier to keep the past behind a locked door and to keep silent on family tragedy.

"I still can't believe something like this happened in my family," Diana told me. "I never got to know my aunt Sally. I also wish I could have been able to support my mom during that awful period."

By this point, the Panaros had had a son, Brian, born in 1968, twenty years after Diana. Susan and Al had given up on the greenhouse well before Brian's birth (a surprise pregnancy after several additional miscarriages made the likelihood of another child seem all the more remote). They had also left New Jersey for South Carolina, with Ella joining them for a time, though they would return to their home state in the mid-1970s. Susan became a full-time homemaker. She spent ample time gardening—for pleasure, not for income—volunteering with her church, and with her family.

When the Panaros returned to New Jersey, Ella settled back in New Egypt, where she had spent much of her childhood and early motherhood. She then moved to a nursing home in Pemberton, where she died in 1998 at the age of ninety-one. Susan died in 2012, and Al passed away in February 2016.

"DID YOU SAY THAT SALLY HORNER was the inspiration for *Lolita?*" Carol Starts said at the beginning of our first conversation in December 2016, once I'd explained why I was calling her out of the blue. Her incredulity was palpable. So, too, was her admiration for her long-deceased best friend.

"I was so unbelievably impressed by her. Sally taught me a great deal. After she was gone, I went through modeling classes to be a 'lady.' Because that's the way Sally was. I wanted to be like her. So I was. I went through those classes, how to walk and sit and stand and so forth. I paid attention to actions, movements, how to dress, and thoroughly enjoyed it because I could be just like Sally. I was on the wrong side of the tracks. I didn't have much mentorship prior to her."

The strong impression Sally made endured for Carol's entire life. She left Camden for California at eighteen to marry
her first husband. She kept his last name of Taylor—she was
glad to shed her maiden name, which had caused her no end
of teasing at school—married and divorced three more times,
and had four children. Carol died on October 30, 2017, in Mel-
bourne, Florida, where she lived with one of her daughters. She
was eighty years old.

EDWARD BAKER GOT ON with his life after the accident that
killed Sally. He was drafted into the army in the summer of
1954 and spent more than eight months at Schweinfurt, Germany, where he celebrated his twenty-third birthday. Upon returning to Vineland, he married, had a son he named Edward
Jr., and worked as a machinist, with side interests in riding
and fixing motorcycles and watching NASCAR. Both father
and son volunteered substantial amounts of time with the local
YMCA. But several years before Edward Baker's death in 2014,
fate had another twist in store for him.

On the afternoon of Wednesday, May 17, 2007, his son fell
asleep in his Mercury Grand Marquis. The car crossed the middle lane and into the median, then careened along the shoulder
of Route 55, slamming into a tree. Just as Sally Horner had,
fifty-five years earlier, Edward Baker, Jr., died instantly.

THE TWO CAMDEN POLICE DETECTIVES involved with Sally
Horner's rescue, Marshall Thompson and Wilfred Dube, both
retired in the mid-1960s. Dube had risen to become chief of
detectives; Thompson remained a detective until his sixtieth

birthday. Dube died in 1980; Thompson in 1982. Howard Hornbuckle served one more term as Santa Clara County sheriff, and retired to work as a dairy farm sales representative. He died in 1962.

MITCHELL COHEN'S HEALTH suffered after the Sally Horner case concluded. Her rescue, and Frank La Salle's imprisonment, took place only a few months after Howard Unruh's massacre, an exhausting one-two combination for the Camden County prosecutor. Cohen spent three days in the hospital at the end of August 1950. Doctors ordered him to take a rest from his work; he took their advice and left Camden for a week in the White Mountains of New Hampshire.

Upon his return, and over the next eight years, there were more major crimes for Cohen to prosecute, including the ones resulting in the 1955 execution of three men for murdering a luncheonette owner during a botched robbery. Then came his next career move, when the new New Jersey governor at the time, Robert B. Meyner, appointed Cohen as Camden County Court judge. In doing so the governor displaced Rocco Palese, the judge who had sentenced Frank La Salle to prison. Controversy ensued in the form of a letter-writing campaign and newspaper editorials, but Palese eventually acquiesced and moved into private practice. (He died in 1987 at the age of ninety-three.)

Cohen served three years on the county court bench before moving over to the appeals court for a year. He was then appointed a federal court judge by President Kennedy in 1962, and he stayed on the bench for the rest of his life. Cohen died in 1991, at eighty-six.

AS *LOLITA* BEGAN its climb up the bestseller charts, Véra
Nabokov continued what had become a custom for her since
May 20, 1958: jotting down her private thoughts in the diary
that had previously been the sole property of her husband.
Véra's notes largely indicate delight at *Lolita*'s success, but one
subject bothered her above all: the way that public reception,
and critical assessments, seemed to forget that there was a little girl at the center of the novel, and that she deserved more
attention and care:

> I wish someone would notice the tender description
> of the child's helplessness, her pathetic dependence
> upon the monstrous HH, and her heartrending courage
> all along, culminating in that squalid but essentially
> pure and healthy marriage, and her letter, and her
> dog. And that terrible expression on her face when
> she had been cheated by HH out of some little plea-
> sure that had been promised. They all miss the fact
> that the "horrid little brat" Lolita is essentially very
> good indeed—or she would not have straightened out
> after being crushed so terribly, and found a decent
> life with poor Dick more to her liking than the other
> kind.

Véra, of course, did not intend her thoughts for publication, and Vladimir Nabokov did not express these thoughts
in public, either. *Lolita*'s success almost seemed designed so
people missed the point. Its original publication by Olympia
Press established its bona fides as a book too controversial for
American consumption. And then, once it was finally published in the United States, the conversation centered around
Humbert Humbert's desires and his "love story" with Dolores

Haze, with few acknowledging, or even comprehending, that their relationship was an abuse of power.

As a result, that left a vacuum for decades of readers to misinterpret *Lolita*. It allowed for a culture of teen-temptress vamping that did not account for the victimization at the novel's core. Sixty years on, many readers still don't see through Humbert Humbert's vile perversions, and still blame Dolores Haze for her behavior, as if she had the will to resist, and chose not to.

LATER IN THE YEAR, Véra wrote in her diary about a strange evening that foreshadowed all the ways in which *Lolita* would be viewed as grim comedy instead of the moral indictment she'd hoped for. On November 26, 1958, Vladimir and Véra Nabokov went out to dinner at Cafe Chambord on Third Avenue and Forty-Ninth Street. The other dinner guests included Walter Minton and his wife, Polly, as well as Victor Schaller, Putnam's head of finance, and his wife. The mood should have been celebratory in light of *Lolita*'s increasing success. It was not, as Véra later wrote in great detail, because the Mintons were unduly preoccupied with a *Time* magazine article published the previous week.

The article, unbylined but written by staff writer and future *Los Angeles Times* gossip columnist Joyce Haber, was ostensibly about the public reception of *Lolita*. Haber opened with an account of Nabokov at a Putnam-sponsored reception for the novel, where he, according to Haber, "faced a formidable force of 1,000 literature-loving women." After quickly dispensing with the positive and negative critical reception for *Lolita*, Haber let loose on Rosemary Ridgewell, the showgirl-turned-literary-scout, with carefully calibrated bile.

Haber described Ridgewell as "a superannuated (27)

nymphet . . . a tall (5 ft. 8 in.) slithery-blithery onetime Latin Quarter showgirl who wears a gold swizzle stick around her neck and a bubbly smile on her face. Well may she bubble." Ridgewell merited Haber's attention for tipping off Minton to *Lolita*'s existence after reading excerpts in the *Anchor Review.* But the cause of Haber's ire was that she and Ridgewell were Walter Minton's mistresses at the same time. No wonder she felt compelled to douse her rival in the prose equivalent of hydrochloric acid.

Véra Nabokov would learn some of these details at the Cafe Chambord dinner, where she sat next to Walter Minton's wife, Polly. The younger woman—"a pretty girl, rather unhappy"— immediately began to unburden herself to Véra, whom she'd never met. The "frightened, bewildered" Polly looked upon *Lolita* as a source of pain and problems in her marriage to Walter. Where once the couple was happy, Polly confided, since the novel's arrival in their lives her husband "began to see a lot of people and get mixed up."

Polly let slip to Véra that she first learned of her husband's involvement with Ridgewell through the "horrid" *Time* article. Véra was, apparently, unnerved by Polly's confession, but had the wherewithal to observe in her diary: "Poor Polly, small-town little girl, craving for so many pounds of 'culture' gift-boxed and tied with a nice pink bow!" Véra did not know Rosemary, but based on what Polly told her and the *Time* article, she judged her as "a pretty awful, vulgar but flashy young female."

Odd as this encounter was for Véra, the evening devolved further. After Victor Schaller and his wife bid the Nabokovs and the Mintons adieu, Dmitri turned up, driving his 1957 MG sports car. Polly, enthralled, requested a ride, and Dmitri obliged. Vladimir and Véra took a cab to their hotel, accompanied by Minton, who proceeded—within earshot of the driver, and perhaps unprompted—to admit his affairs with both Ridgewell and Haber.

"Between his two little harlots," Véra wrote, "M[inton] ruined his family life." Minton swore both affairs were over, that he had "made it up to Polly," and presented Rosemary "in a very unsavory light, a little courtesan, almost a 'call girl,' trying to collect as much money as she could from Walter and spouting nonsense about *Lolita*."

When the trio arrived at the hotel, Polly and Dmitri were still MIA. The Nabokovs and Minton "waited and waited," Véra recording this phrase and then crossing it out. When the duo finally appeared in the hotel lobby, Dmitri informed his parents "with a sly smile" that he and Polly had driven to his apartment, because she had wished to see it. The next day, Véra wrote, "Minton told V., 'I hear Dmitri gave Polly a good time last night.'" Véra did not know what to make of Minton's comment. "I wonder if this sort of thing is normal or typical of today's America? A bad novel by some O'Hara or Cozens [*sic*] suddenly come to life."

The dark comedy of the evening did indeed resemble a John O'Hara story or James Gould Cozens's *By Love Possessed,* which had been a bestseller the year before. What Véra Nabokov witnessed, and grew so disturbed by that she was compelled to write about it in her diary, seemed like a harbinger of all the ways in which American culture would corrupt *Lolita* and misunderstand Nabokov's meaning. If those closest to the Nabokovs were behaving strangely, who else might this novel have the power to corrupt?

LOLITA WAS A PROPER HIT. The more the novel sold, the more people ventured an opinion, whether they had read it or not.

Comedians turned *Lolita* into late-night fodder (Groucho Marx: "I'll put off reading *Lolita* for six more years until she turns eighteen"; Milton Berle: "First of all, let me congratulate Lolita now. She is thirteen"), another signal of *Lolita* reaching a level of success far beyond literary spheres.

Nabokov enjoyed the attention. He gave interviews to journalists and appeared on talk shows on both sides of the Atlantic. A cartoon featuring *Lolita* in the July 1959 issue of *Playboy* amused him enough that he mentioned it to his American publisher, and Véra noted in her diary how she and Vladimir delighted in the jokes broadcast on television.

Lolita's appeal extended to fashion magazines and film, with dissonant, even bizarre results. These depictions were largely knowing, winking parodies, playing up the overt sexuality of certain blond bombshell personas in the guise of younger girls. The most blatant reference to Nabokov's creation, equal parts amusing and disturbing, appeared in the film *Let's Make Love,* which features Marilyn Monroe singing a version of Cole Porter's "My Heart Belongs to Daddy" after announcing: "My name . . . is Lolita. And I'm . . . not supposed to . . . play . . . with boys!"

Another bizarre stunt affected the Nabokovs more personally. Their son, Dmitri, had moved to Milan to pursue an opera-singing career. But it was getting access to his famous father that left Dmitri open to strange requests. A local magazine covinced him to judge a contest where the winner would pose as Lolita for a fashion shoot to be held at his own apartment. Dmitri, reflecting on his youthful stupidity, recalled that the "decidedly post-pubescent aspiring nymphets, some with provincial mothers in tow," had invaded his apartment for two solid days.

Newspaper coverage of the contest reached Dmitri's father,

who was upset enough to send a telegram to his son asking for the contest to be stopped. "The publicity is in very bad taste," Nabokov wrote Dmitri on October 7, 1960. "It can only harm you in the eyes of those who take music seriously. It has already harmed me: because of it I cannot come to Italy since the reporters would immediately pounce on me there." Nabokov was especially disappointed in Dmitri for letting "this unhealthy ruckus" overshadow his own career. Dmitri learned his lesson. From that point on, he would defend *Lolita*'s honor rather than corrupt it. But the contest mess was further proof of the ways in which perceptions of *Lolita* moved from tragedy to carnival.

BY THIS POINT Nabokov had completed a screenplay draft of *Lolita* for Stanley Kubrick. Nabokov had initially turned it down—"by nature I am no dramatist, I am not even a hack scenarist"—but while on an extended European vacation at the end of 1959 and the beginning of 1960, he relented. On January 28, temporarily ensconced in Menton, France, Nabokov wrote his friend Morris Bishop that he changed his mind about adapting *Lolita* because "a pleasing and elegant solution of the problems involved suddenly dawned on me in the gardens of Taormina."

Contract from Kubrick and his producing partner, James Harris, in hand, the Nabokovs ventured west to Los Angeles, arriving in March. Vladimir holed up for the next few months to complete the screenplay. The first draft, finished in August 1960, ran more than four hundred pages long. That draft was not used, and neither were subsequent ones Nabokov wrote before he and Véra sailed back to Europe in November. Kubrick rewrote the screenplay substantially before shooting the film the following year, though Nabokov was still given sole screenplay credit when the film was released in 1962 and he was subsequently nominated for an Academy Award.

What most surprised me about the original *Lolita* screenplay draft, which I read at the Berg archives of the New York Public Library, were two names that appeared in a second-half scene that did not survive in the film version, or in Nabokov's published screenplay in 1973. The names are perhaps coincidental, but they didn't seem that way to me. It felt more like unfinished business; that Nabokov was not through mining Sally's kidnapping for his creative pursuits.

In the scene, Humbert Humbert mentions a "Gabriel Goff," who is the subject of a gala in Elphinstone. "Goff, a blackbearded railway robber, held up his last train in 1888, not to rob it but to kidnap a theatrical company for his and his gang's entertainment. The stones in Elphinstone are full of Goff faces, bearded pink masks, and all the men have grown more or less luxuriant whiskers." Goff happened to be the maiden name of Sally Horner's mother, Ella.

Later in the scene, Nabokov makes a repeated reference to a "Dr. Fogg," a doctor deemed to be the best to treat Dolores's illness. It turns out "Fogg" is a disguise for Clare Quilty, who uses that alias but arrives on the scene wearing the mask of the railway robber, Gabriel Goff. Fogg, of course, was an early alias of Frank La Salle.

I could chalk up the use of these names to Nabokov's merry-trickster side, noting that "Goff" and "Fogg" are inversions of each other. The presence of doubles and masks certainly bolsters this theory. But because the Sally Horner parenthetical reference was excised from the film script—there was no reason to preserve a textual reference for a visual medium, after all—this name-inversion trick read to me as if Nabokov wanted to preserve the link to Sally's story in some fashion.

JAMES MASON SIGNED ON as Humbert Humbert, while Peter Sellers and Shelley Winters were cast as Clare Quilty and Charlotte Haze, respectively. The final piece of the *Lolita* film puzzle was choosing the girl to play Dolores Haze, a search avidly covered by newspapers and magazines around the world. One gossip item speculated that Mason's eleven-year-old daughter, Portland, might be cast. Tuesday Weld, already an established television and film star at seventeen, was a serious contender, much to Nabokov's chagrin ("a graceful ingenue but not my idea of Lolita"), but she declined the role, famously saying, "I didn't have to play Lolita. I *was* Lolita."

When Sue Lyon, age fourteen, was ultimately cast, it was with Nabokov's enthusiastic approval. He did not want a girl so near to Lolita's true age for the film, and he agreed with Kubrick that a girl who looked closer to sixteen, as Lyon did, would circumvent the censors. *Lolita* could not go forward with an "X" rating, or worse, no rating at all. When Lyon began shooting *Lolita* in 1961, European newspapers followed her around taking photographs of her on set, buying food, or taking a nap. Typical paparazzi behavior, but it seemed that much more invasive because they were chasing a teenager playing the love interest of a much older man.

However Kubrick, as director, and Nabokov, as author, envisioned the reception of *Lolita* the film upon its release in the summer of 1962, they were disappointed. The infamous Bert Stern photograph of Lyon, sucking on a lollipop and wearing heart-shaped glasses, shaped public perception. So did the tagline "How Did They Ever Make a Movie of *Lolita*?" which caused a number of critics to answer: they didn't. Lyon was too old, unconvincing save for the scenes where she has transformed into the pregnant Mrs. Dick Schiller. Kubrick blamed the censors for his creative misfire. Nabokov, more generously,

judged Kubrick's vision "a first-rate film with magnificent actors."
While the movie did all right at the box office—$9.25 million
grossed on a $2 million budget—the critical reception cast a pall.

As time went on, *Lolita* was adapted repeatedly—again as a
1997 film, as a 1981 play by Edward Albee, as a 1990s Russian-
language opera, and even as a musical. The history of these
adaptations, nearly all by middle-aged men, indicate how far
out of touch they were from the novel's core depiction of sexual
abuse. Reading *Lolita* allowed people to make their own judgments, rightly or wrongly. Seeing or hearing her sing, dance,
or speak provoked far more uncomfortable responses, which
led to a host of failed projects. That anyone, let alone those with
a financial stake, could think visual and theatrical depictions of
Lolita would be successful seems laughable in hindsight.

The most ludicrous idea was *Lolita, My Love,* the 1971 musical version. And yet it boasted an A-list group of creators, with
lyrics and libretto by Alan Jay Lerner, who wrote the smash
hit *My Fair Lady,* and a score by John Barry (of James Bond
theme fame). Nabokov, who could hardly abide music, gave his
approval for the musical because even he was aware of Lerner's
and Barry's past successes. But the result never reached Broadway. Savage reviews of the original Philadelphia production
closed it down in early 1971, and a revived version staged in
Boston a couple of months later—starring future *Willy Wonka
& the Chocolate Factory* star Denise Nickerson, not yet thirteen,
as Dolores—yielded more disappointing notices. Nabokov had
once told a disapproving interviewer that *Lolita, My Love* was
"in the best of hands." Once the musical failed, he never spoke
of it again.

Lolita also spawned unauthorized sequels that, in different
ways, demonstrated Nabokov's mastery of difficult material,
which suffered in the hands of far less talented writers. *The*

Lolita Complex, published in 1966 by an ex-con named Russell Trainer, purported to "investigate the activities of real-life Lolitas and Humberts and offers insights into an important social problem" through "case histories, professional opinions, court transcripts, interviews and police records." Trainer even thanked several of these medical professionals by name, but I could not verify that any of them existed. They are likely as fictitious as Nabokov's invented John Ray, Jr., who supplied the parodic introduction to Humbert Humbert's memoirs. Nabokov not only drew from Havelock Ellis's history of sexual deviants, but also reacted to the pervasive influence of Sigmund Freud—whose psychoanalytic theories he detested. "I think he's crude. I think he's medieval, and I don't want an elderly gentleman from Vienna with an umbrella inflicting his dreams upon me," Nabokov huffed in a 1965 interview.

The Lolita Complex was a crude cash-in, written by a veteran writer from the paperback porn mills who began his career while in prison for check fraud. Trainer's book did enough business for him to write a 1969 sequel, *The Male Lolita,* in which the faux–case history format shifted focus to young boys in power-imbalanced relationships with women. And Trainer's literary contributions might have stayed forgotten save for an improbable twist: in its Japanese translation, *The Lolita Complex* became a foundational text for the development of manga and anime, particularly the "lolicon" subgenre where little girls with big doe eyes are depicted as objects of desire and in explicit sexual situations. ("Lolicon" is a portmanteau of "Lolita Complex.")

Thirty years after *The Lolita Complex,* another unauthorized sequel took a different approach, retelling *Lolita* from Dolores Haze's perspective. *Lo's Diary,* by the Italian journalist Pia Pera, proved to be a missed opportunity. Instead of getting

at the truth of Dolores Haze's dark plight, of showing her the
way even Nabokov hinted at—as a clear victim, struggling to
survive and maintain some sort of agency when she could
never have enough power—Pera's version of Lolita depicted her
as a brazen seductress, her behavior more reminiscent of Veda,
the young (but not underage) daughter in James M. Cain's *Mildred Pierce*. *Lo's Diary* also suffered from years of litigation with
the Nabokov estate, which blocked its publication in English
until 1999.

Two years earlier, Adrian Lyne's film remake of *Lolita* arrived out of its own legal quagmire, having faced almost as
many censorship issues as did Stanley Kubrick's. Lyne's film,
scripted by Stephen Schiff, is quite faithful to Nabokov's novel.
Jeremy Irons is almost too perfectly cast as Humbert Humbert
(he later lent his voice to the audiobook edition of the novel
issued on *Lolita*'s fiftieth anniversary). Dominique Swain is
starkly believable as Dolores, holding her own against Irons's
all-encompassing talent, and Frank Langella shines as Clare
Quilty.

The cultural climate had shifted back and forth between
liberal progressiveness and conservative backlash in the intervening thirty-five years, but in 1997 the appetite for a new film
version was particularly low. Lyne had tried and failed to make
the film for years. Once he had finally finished shooting, he
faced fresh legal issues, after the passage of the Child Pornography Prevention Act of 1996, which made illegal any visual
depictions of children having sex with adults—whether or not
a child was involved.

Lyne battled lawyers seeking significant cuts to the film
and struggled to find a distributor, which delayed *Lolita*'s opening in North American theaters by more than a year. The theatrical run was tiny (to qualify for the Academy Awards) and

a prelude to an airing on the cable television network Show-time. This *Lolita*, as a result, did even poorer box office business than its predecessor. Once more, the general public did not have much appetite for seeing *Lolita* on-screen, as opposed to imagining her within the covers of a book.

More than sixty years on, the appetite for adapting *Lolita* or reviving earlier adaptations has likely subsided for good. It is difficult to see how it could be done, especially given the growing polarization of the political climate. The dark heart of *Lolita*, and the tragedy of Dolores Haze, may now be too much to transform into entertainment. It's wiser, and saner, to remember the little girl at the center of the novel, and all of the real girls, like Sally Horner, who suffered and survived.

On Two Girls Named
Lolita and Sally

B oth times I met with Sally Horner's niece, Diana Chiemingo, she picked me up at the Burlington Towne Center light rail station in New Jersey and drove us a mile down the road to Amy's Omelette House, which does, in fact, specialize in omelets. Diana turned seventy in August 2018. Her figure is slight and her voice does not carry, but both convey a steely toughness. She gets to a point quickly and is not prone to running on. Silences often stretched between us as she considered how to phrase her answers in just the right way.

Sally is never far from her niece's thoughts. It was particularly apparent in our first face-to-face meeting in the summer of 2016. Each of us arrived at the diner with photos to show the other. Diana brought a stack of black-and-white images of Sally, Susan, Al, Ella, and others—her best friend, Carol, Sally's un-identified date for the evening social, probable classmates at Burrough Junior High, possible acquaintances from her last summer. Both Diana and I marveled at what a fully grown, vibrant girl Sally appeared to be. A vibrancy that had so little time to assert itself.

Then it was my turn. My photos—grainy images scanned from the *Courier-Post* coverage of Sally's rescue—were nowhere near in as good condition as Diana's trove. But I knew she needed to see the one of her, not quite two, sitting with her parents as Susan speaks to Sally on the telephone, hours after her younger sister's rescue in San Jose. Diana was startled by the photograph—she had never seen it before. Seeing her and her family all together, so long ago, made Sally's story feel fresher, more vivid. The tragic parts, but also the happier parts. Sally had come home and was part of their family again, even if it wasn't for very long.

Diana spoke of her parents, of her grandmother Ella, of her own life. The family tried to hold Sally's abrupt loss at bay with mixed results. She became, and remained, the family phan-tom. For a long time, Diana had no inkling of the *Lolita* con-nection. She'd never read the novel, so of course she would not have seen the reference to her aunt in the text. She learned of the connection when her brother, Brian, a police department evidence technician in Florence, searched online and discov-ered Sally's sparse Wikipedia entry as well as the essay by Al-exander Dolinin.

"He was shocked," Diana told me. "I was, too. I don't know how to explain it. To think that someone is writing about your family? I was so young when everything happened, and for people to be writing about Sally—that's a really big thing."

By our second face-to-face meeting, nearly a year later, Diana had had more time to sit with the idea that Sally's story was part of a larger mosaic of girls and women who had been cruelly wronged and abused by men. Stigmas take a long time to fade. But the more Diana talked about her aunt, the more the relief, and even the joy, showed through to compensate for what she, her family, and Sally had lost.

LOLITA'S POST-PUBLICATION afterlife meant that years later, Vladimir Nabokov was still being asked about the novel, again and again, in interviews. He did not like this. The irritation is evident, leading to contradictory responses about what influenced him. He denied Humbert Humbert had a real-life basis, despite his repeated chess matches with Henry Lanz at Stanford, or his reading of Havelock Ellis: "He's a man I devised, a man with an obsession, and I think many of my characters have sudden obsessions, different kinds of obsessions; but he never existed." He denied Lolita was based upon a real girl, despite the parenthetical mention of Sally Horner. He denied any moral agenda, telling the *Paris Review:* "it is not *my* sense of the immorality of the . . . relationship that is strong; it is Humbert's sense. *He* cares, I do not. *I* do not give a damn for public morals, in America or elsewhere."

To admit he pilfered from a true story would be, in Nabokov's mind, to take away from the power of his narrative. To diminish the authority of his own art. The controlled nature of these interactions, with questions submitted in advance and responses edited after the fact, still left room for surprises—all the more because, as was customary with Nabokov, of what he chose not to say, as well as his exact phrasing of things he did say.

After one stern denial in a 1962 interview, Nabokov changed his tune a little in the near-next breath, saying that Humbert *did* exist, but only after he had written *Lolita*. "While I was writing the book, here and there in a newspaper I would read all sorts of accounts about elderly gentlemen who pursued little girls: a kind of interesting coincidence but that's about all."

This is a close-to-tacit admission by Nabokov that he knew of actual cases that bore some resemblance to his fictional world. Cases like Sally Horner's kidnapping at the hands of Frank La Salle. Nabokov references them in the text of *Lolita*, but to do so in an interview was anathema, lest listeners or readers connect the dots.

But there is no getting around the fact that Nabokov kept returning to this taboo relationship between a young girl and an older man throughout his career. That compulsion had real-life basis not only in other people's lives, but also in his own. A clue to that compulsion emerges when Humbert Humbert describes his "rather repulsive" uncle, Gustave Trapp. He thinks of Trapp again while tracking Ivor Quilty, Clare's dentist uncle, and again during his first encounter with Clare at the Enchanted Hunters hotel. (Quilty also punches through the proverbial fourth wall by signing his name as "G. Trapp" in the hotel's guestbook entry. As the German scholar Michael Maar points out, "Quilty cannot know the name.")

That uncles figure so much in *Lolita* recalls a revelation in *Speak, Memory*: that Vladimir's uncle Ruka took his then-nine-year-old nephew onto his knee and fondled him repeatedly "with crooning sounds and fancy endearments" until the boy's father called for him from the veranda. The real-life scene seems to foreshadow the famous fictional one of Humbert achieving orgasm with Dolores on his lap. Humbert, of course, believes his emission to be furtive—that the girl doesn't know.

But Nabokov, in his description, leaves it up to the reader to decide what Lolita knew.

IN *READING LOLITA IN TEHRAN*, Azar Nafisi makes the excellent point that Dolores Haze is a double victim, because not only her life is taken from her, but also her life story: "The desperate truth of *Lolita*'s story is *not* the rape of a twelve-year-old by a dirty old man but the *confiscation of one individual's life by another.*" Without realizing it, Nafisi has made the exact parallel between Dolores Haze and Sally Horner. For Sally's life, too, was forever marked by the twenty-one months she spent as Frank La Salle's captive, his false daughter, his own realized fantasy. After she was rescued, she attempted to resume the life snatched away from her. And it seemed she did, on the surface.

But how could she, when her story had been front-page news all across the country, and when those in Camden knew exactly what had happened to her and judged her—blamed her—for it? Whether she'd lived two years or many decades, whether she might have had time to move forward, even if she could not move on, Sally Horner was forever marked.

Lolita's end, dying in childbirth, is a tragedy. But Sally Horner's demise by car accident is the bigger tragedy, because it was real, and robbed her of the chance to grow up and at least attempt to move forward. In fact, Sally Horner is a triple victim: snatched from her ordinary life by Frank La Salle, only for her life to be cut short by car accident, and then strip-mined to produce the bones of *Lolita*, the only acknowledgment a parenthetical reference hidden in plain sight, hardly noticed by many millions of readers.

Over the course of researching this book these last few

years, I would ask faithful fans of *Lolita* if they'd caught the parenthetical reference to Sally Horner's kidnapping. The unanimous answer was "no." This was no real surprise. If no one caught the reference, how could they be expected to see how much of the novel's structure rides on what happened to Sally in real life? But once seen, it is impossible to unsee.

There is no simple lock-and-key metaphor to equate the tragic story of Dolores Haze to the tragic story of Sally Horner. Vladimir Nabokov was too shrewd to create a life-meets-art dynamic. But Sally's story is certainly *one* of those important keys that, once employed, unlocks a critical inspiration. There is no question *Lolita* would have existed without Sally Horner because Nabokov spent over twenty years dwelling on the theme, working it out in bits and pieces as he moved around Europe and America. But the narrative was also strengthened and sharpened by the inclusion of her story.

Sally Horner can't be cast aside so easily. She must be remembered as more than a young girl forever changed by a middle-aged man's crime of monstrous perversion. A girl who survived adversity, manipulation, and cross-country horror, only to be denied the chance to grow up. A girl immortalized, and forever trapped, in the pages of a classic novel of satire and sadness, like a butterfly with wings damaged before ever having the chance to fly.

Sally Horner, age fifteen, summer of 1952.

AFTERWORD

On a Friday morning in August 2018, just over two weeks before the publication of *The Real Lolita*, an email from the FBI landed in my inbox. The message was quite bland—"There are eFOIA files available for you to download"—with two links attached. I had forty-eight hours to download the files, after which the links would no longer work.

I'd requested these files through the federal Freedom of Information Act more than a year earlier, and had long given up on a response. My book was finished, edited, ready to print. There were missing pieces, information I longed to have, about Sally Horner's kidnapping and Frank La Salle's early life, but I'd accepted I might not discover them, and wrote around them. The FBI operated on its own schedule, according to its own whims, caring little as to whether they might overlap with my own.

After downloading the files, I opened the PDFs, which together comprised sixty-eight pages of Frank La Salle's FBI file (another forty pages were redacted entirely). Reading through the pages, which began with an August 1948 teletype about Sally being reported missing by her mother, Ella, and ended with La Salle's April 1950 guilty plea, I gleaned some of the information I craved all this time.

One throwaway line in particular turned out to unlock a

riddle I'd spent more than four years trying to solve. I wanted
to know La Salle's whereabouts before 1937, when he met and
married Dorothy Dare, his second wife, but leads kept disappearing into dust. One repeated exercise in frustration centered around La Salle's time in Leavenworth State Prison in
the mid-1920s. The trouble was, as I wrote in the book, La
Salle used so many aliases—I would eventually tally more than
twenty—that finding the correct pseudonym was a challenge.
None of the aliases I knew of checked out. Not Frank LaPlante,
not Frank Warner, not Jack O'Keefe, not Frank (or Harry) Patterson.

Thanks to his FBI file, I finally had the right fake name:
Frank Campbell. I also finally had his inmate number at Leavenworth: 22217. Alighting upon both, and as quickly as I could,
I sent off an email to Greg Bognich, the archives technician at
the National Archives office in Kansas City. Over the years he
had shared in my frustration that we could not confirm Frank
La Salle served prison time in Leavenworth. This time, however, Bognich's response, sent two and a half hours after my
email, was quite different:

Sarah! OMG!

*I can't believe you were able to finally find this guy! This
one had me stumped for the longest time. I looked in his file
and it is indeed Mr. La Salle.*

When the full Leavenworth file arrived in my inbox a few
days later, I knew exactly why Bognich could be so certain: its
first page was a prison intake photo. The man was clearly a
younger Frank La Salle.

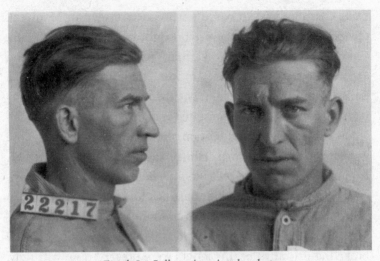

Frank La Salle, prison intake photo,
Leavenworth, Kansas, December 1924

This file included not only a photo, but also a full accounting of La Salle's time in Leavenworth.*There were intake forms describing the nature of his crimes, stemming from running a car theft ring in Indiana and Pennsylvania and violating the Dyer Act, which forbid transporting stolen goods across state lines; and the length of his federal prison sentence, which began in December 1924 and ended in January 1928, after an early parole for good behavior.

There were detailed notes listing every malady and remedy, every work detail, and every violation and removal of privileges. There were biographical details that had some ring of truth—all my investigative efforts settled on La Salle emigrating to the United States as a young boy, likely under the age of five,

---

* Chapter seven, "Frank in Shadow," has been revised to reflect this full accounting and new details about La Salle's early life.

from Canada, born to parents who once resided in Montreal, Quebec—and some that were outright fabrications. As Frank Campbell, he purported to be married with five children, and around the age of forty-five. At most, from the intake photo, La Salle looked to be about thirty, albeit a weather-beaten thirty with a clear affinity for cigarettes and alcohol.

What really got my attention, though, were the letters. One was on behalf of a lovelorn cellmate (using the pseudonym A. P. Anderson), and another was to a former lover he had no business writing to because their parting had been acrimonious. It was against Leavenworth rules to forge another man's signature or to contact those who expressly refused it. Since La Salle broke the prison rules, his letters were preserved, unwittingly, for posterity. That meant I had samples of La Salle's handwriting. I'd never seen his written words before, other than signatures, because at Trenton State Prison, where he spent the last sixteen years of his life, La Salle used a typewriter.

It was disorienting and thrilling to read his letters, some to prospective or previous partners in criminal trade, and others to young women La Salle had once known and wished to get back in their good graces. A letter La Salle, as Campbell, sent on December 22, 1925, to a woman named Olin Frye, residing in New Stanton, Pennsylvania, was addressed "to my former dear friend." After wishing her a Merry Christmas and Happy New Year, he got at the reason for writing: "No doubt you are still very mad at me which I don't blame you for being so, considering the facts as they were at the time."

La Salle wanted a favor: a photograph of one Lady Mae, "which I value very highly, and if you would please send it to me as it is the last of those pictures and I always wished to keep that one as a remembrance to a true and real little Pal I once had." In exchange, he wrote, he'd send Olin "a real Bead Mesh

Hand Bag worth Thirty or Forty Dollars," and even get one to "your Dear Mother or Becke or Dorthy" for how kindly they treated him in the past. La Salle also implored Olin "to please tell little Walter his big old pal still thinks and remembers him and shall always do so."

He did not, however, tell Olin Frye that he was in prison. Instead, La Salle made up an entire story about expecting "in a couple of years to go over to Europe for a very responsible firm in La Havre [sic], France" that "made me a real good yearly rate," which he planned to accept (spelled "exept"). After more lies about prospects in South America, he signed off: "Your Old Friend of the Past, Mr. Frank Campbell."

The letters wheedled and cajoled, flattered and made demands. They lacked humor. At times they were downright creepy. The closest any came to predicting his future predilections for young girls was an undated missive to a woman named Florence Yohey, who lived in Philadelphia and clearly had once had a relationship with La Salle. He referred to himself in the letter as "daddy" and spoke of wanting her to sit on his lap. (He also signed that one with the A. P. Anderson pseudonym.)

La Salle's contraband Leavenworth correspondence likely contributed to his parole denial in January 1926. And they certainly restricted his ability to write letters to his mother— mostly to alert her that he could not, in fact, attend the funeral of his stepfather—and a sister. Two years later, on January 7, 1928, Frank Campbell was released from prison. He went back to Philadelphia and disappeared into the void for the next decade.

A document trove answers as many questions as it opens up avenues for new ones. Now I knew what Frank La Salle looked like as a younger man, and where he was originally

from, but so much information about him eluded my grasp.
When, precisely, did he and his family emigrate to the United
States? Did he live in Chicago, or was that only his mother, and
perhaps a sister? Did La Salle serve in World War I? When did
he start spending time in Indianapolis? Who was Olive Lakin,
the young woman La Salle was arrested with in Greensburg,
Pennsylvania, in August 1923, and was she the "Ollie" whom
he said was his first wife? Was James W. Hensley, his apparent partner in the car theft ring, the same Hensley who served
shockingly little time fifteen years earlier for a double murder
stemming from a love affair gone awry?

    And going far beyond facts, could the early behavior and
habits of Frank La Salle shed real light on how he ended up
molesting so many girls and kidnapping Sally Horner?

    Once more, I would have to be content with knowing some
of the answers, but not nearly enough of them. Frank La Salle
was less of a mystery now, cemented in my mind as a common
criminal turned opportunistic child molester. But I hadn't
solved him outright. And it's likely I never will.

THE FBI FILE began with a memo on August 10, 1948, from
J. Edgar Hoover to Olin Jessup, the Special Agent in Charge
at the Bureau's Newark department, noting the recent news
reports about Sally Horner's kidnapping and Frank La Salle's
status as a "convicted rapist reportedly taking [the] victim to
Baltimore, Maryland." The memo asked the Newark office to
look into the matter and see whether there was a "federal violation" involved. Newark responded two days later, on August
12, with a two-page memo outlining the general gist of what
Ella told the Camden police, though they had different details.

In their writeup, the alias La Salle used was Jack O'Keefe, not Frank Warner; the memo also claimed Ella received an earlier letter from Sally, stating that she and La Salle were headed to New York, before switching the expected destination to Baltimore in a second letter, "written in the daughter's own handwriting."

The tone of the memo made clear that the FBI wasn't going to take Sally's kidnapping all that seriously, though. Jessup, the special agent in charge, discussed the matter with the area federal prosecutor, Grover C. Richman, who "advised [he] would not prosecute [based on] info available." However, Richman suggested that "if it developed that subject transported victim across state line for immoral purpose he would then consider prosecution."

On August 20, 1948, Ella Horner went to the Philadelphia office of the FBI, with Richman also present, to relate her account of Sally's disappearance. Once more, the AUSA made clear his misgivings that there was anything to prosecute: "There is nothing to indicate any internal state transportation or transportation for immoral purposes." Richman also indicated that when Sally returned home—because, at that point, they clearly believed she would—"if she furnished evidence of interstate transportation and immoral acts he would consider prosecution."

Less than two months later, the feds were completely out of the case. As the Newark SAC wrote headquarters on October 13, 1948, "The Camden, New Jersey police department has heard nothing further from the Victim . . . and there has been no showing of interstate transportation to date." Richman further advised that no federal statute had been violated, but again reiterated that if Sally returned home and "the investigation discloses that she has been transported interstate, the Camden, New Jersey police department will notify this office and the Victim will be interviewed."

With that edict, "the case is being considered closed." The FBI ceased to investigate any further. Sally Horner, meanwhile, was in Baltimore, two months into living on 437 East Twentieth Street, and weeks into attending St. Ann's School on Greenmount Avenue under an alias closely linked to one of Frank La Salle's most common aliases. She hid in plain sight from federal law enforcement that didn't bother to look for her.

The FBI never acknowledged their premature closing of the case, and the long delay in bringing Sally Horner home, to the public. But a memo from Special Agent in Charge Al Rosen to his supervisor, Mickey Ladd, did discuss it. Rosen's memo, sent on March 22, 1950, the day after Sally Horner's rescue from the San Jose trailer park, noted, underlining the words for emphasis, that the case had been closed by Grover Richman in October 1948. Rosen handwrote an extra missive in the margins: "It is inconceivable that US Atty [sic] would so rule after [Sally] disappeared for 4 months at 11 years old and taken from a city and a state. It certainly was bad."

THE FBI'S INDIFFERENCE to Sally Horner's plight didn't surprise me. It was part of a larger pattern of indifference she faced during her life and in the many decades following her premature death. The slap of fury, reading Agent Rosen's note, didn't sting any less. Closing a case where there was clear evidence that a kidnapped girl had been taken across state lines by a convicted child rapist? Yes, it certainly was bad.

The Real Lolita's publication showed me, though, how many readers were the opposite of indifferent, that they cared about what happened to Sally. How they wished to know more about her, the real girl, not just the archetype of a kidnapped and rescued victim, or inspiration for Vladimir Nabokov's novel. They were readers who had never heard of *Lolita* or weren't certain if they wished to read the novel, and readers well-versed in all things Nabokovian. I heard from them at events across the United States and Canada. They wrote me emails or sent messages via social media. I am grateful to all of them.

The opinions of two people mattered most to me, and made book events in Philadelphia and Portland, Oregon, stand out. I believe, and said often, that Sally Horner may belong to the world now, but she still has family. People whom she loved, and who loved her in return. Who still miss her dearly, even those born after her death.

When Diana Chiemingo, Sally's niece, attended my Philadelphia event along with her husband and younger brother, I knew it would be an emotionally taxing experience. I sensed her reactions out of the corner of my eye as I spoke of Sally's ordeal, of the car accident that killed her, and of how her life was co-opted by *Lolita*. After the event, while signing a copy of my book for another relative—I'd given Diana a copy of her own a month before *The Real Lolita*'s publication—I asked if my talk had been difficult for her to witness.

"You might have heard a sniffle every now and then," replied Diana.

I'd had some advance warning of Diana's attendance in Philadelphia. In Portland, speaking to a sympathetic crowd at the city's annual book festival, I was taken aback to see a woman waiting for me at the signing table when I arrived there. It was Rachel, Ruth Janisch's daughter, who had devoted so much time and energy speaking of her mother's deeply complex and fraught life, the damage done to her and her siblings, and the one decent thing Ruth did: rescuing Sally Horner from Frank La Salle in San Jose.

Rachel was there with a friend. They had sat through the entire event. The day before, I'd read the chapter about Ruth while standing in front of an Edward Hopper painting at the Portland Art Museum. Had I conjured Rachel up? The last time we'd spoken, she had read through parts of *The Real Lolita* and confessed it to be a difficult read, but had still given her blessing to the text. Did it hold?

Without thinking, after greeting her, I moved in for an embrace. Rachel returned the hug. She told me she was glad to be there. Her friend bought an extra copy of *The Real Lolita*.

Writing a book, even a book that requires interviews with many people, is a solitary process. Publication is a communal effort. A year later, so many more have invested their hearts and minds in learning more about Sally Horner. It's never possible to breathe full life into a dead girl, but now more people are aware, and want to know, how much Sally mattered, and still does.

ACKNOWLEDGMENTS

The Real Lolita has a single author—me—but could not have been written or published without the input, advice, support, and sounding board of many people, in ways large and small. This odyssey began when Jordan Ginsberg, editor-in-chief at *Hazlitt*, replied to my March 2014 article pitch about Sally Horner's kidnapping: "Just brought this up in our editorial meeting, and it got one of the fastest and most enthusiastic 'yes' votes I've heard in a while." Eight months later, in great part to Jordan's editorial vision, the piece was published and changed the course of my professional life. Much has transpired in the intervening four years, and I remain thrilled that it all began at *Hazlitt*. Additional thanks to senior editor Haley Cullingham, whom I have loved working with and hope to do so again soon.

Transforming Sally Horner's story from a magazine piece to a book was equal parts challenging, exhilarating, exhausting, and rewarding. Shana Cohen offered invaluable feedback on the first rounds of book proposal drafts. My agent, David Patterson, has been a brilliant advocate and champion of this project, as has the entire team at the Stuart Krichevsky Literary Agency, particularly Aemilia Phillips, Hannah Schwartz, Ross Harris, and Stuart Krichevsky. Thanks also to my UK agent, Jane Finigan at Lutyens & Rubinstein.

My wonderful editors, Zack Wagman at Ecco and Anne Collins at Knopf Canada, pushed me to meet my ambitions for *The Real Lolita* and then exceed them. I am fortunate to have had such incisive and thoughtful editorial guidance from two of the very best in the business. And to Holly Harley, my editor at Weidenfeld & Nicolson in the UK, thank you for your continued support and never wavering in your enthusiasm for the project.

At Ecco, thanks to Miriam Parker, Sonya Cheuse, Meghan Deans, Megan Lynch, Denise Oswald, Dan Halpern, James Faccinto, Ashley Garland, Martin Wilson, Sara Wood (for the heart-stopping cover design), Allison Saltzman, Lisa Silverman, Andrea Molitor, and especially to Emma Janaskie. At Penguin Random House Canada, thanks to Sarah Jackson, Pamela Murray, Max Arambulo, Marion Garner, Matthew Sibiga, Sarah Smith-Eivemark, Liz Lee, Jared Bland, Robert Wheaton, and Kristin Cochrane.

Special thanks to the MacDowell Colony, for the gift of time and space to finish the first draft of the book; to Karen Riedenburg and David Dean, for invaluable research assistance; to all the archivists and institutions I visited for my research, and the sources who were generous with their time and interviews (more on them in the Notes section); and to Diana Chiemingo, who gave me her trust, faith, and belief that I could do full justice to the brief life of her aunt Sally.

Thank you to friends, family, and colleagues, a list that is by no means comprehensive: Megan Abbott, Jami Attenberg, Alice and Julian AvRutick, Louis AvRutick, Dov Berger, Liza Birkenmeier, Taffy Brodesser-Akner, Michael Cader, Steph Cha, Pamela Colloff, Julia Dahl, Hilary Davidson, Michelle Dean, Robin Dellabough, Nina Elkin, Lyndsay Faye, Dedi Felman, Charles Finch, Jordan Foster, Emily Giglierano, Juliet

Grames, David Grann, Peggy Hageman, Reyhan Harmanci, Lauren Milne Henderson, Ella Hickson, Cara Hoffman, Elizabeth Howard, Janet Hutchings, Hillel Italie, Ethan Iverson, Maureen Johnson, Rokhl Kafrissen, Stephen Karam, Leslie Kauffman, Bob Kolker, Scaachi Koul, Sara Kramer, Maris Kreizman, Clair Lamb, Michelle Legro, Katia Lief, Laura Lippman, Mimi Lipson, Lisa Lutz, Michael Macrone, Jeffrey Marks, Laura Marsh, Kyla Marshell, Chantelle Osman, Helen Oyeyemi, Bud Parr, Andrea Pitzer, Bryon Quertermous, Naben Ruthnum, Alex Segura, Deb Shoval, Kathy Smith, Erin Somers, Daniel Stashower, Adam Sternbergh, Sara Stopek, Caryn Sweeney, Vu Tran, Sharon AvRutick Wallace, Joe Wallace, Robin Wasserman, Deborah Wassertzug, Dave White, Alina Wickham, and Jennifer Young.

Lastly, thank you to my brother, Jaime; the memory of my father, Jack, who I know would have been prouder than anyone that I published this book. And to my mother, Judith, forever my hero.

NOTES

This book is based extensively on primary sources wherever available, including court documents and transcripts, prison records, legislative records, and testimony. I am grateful for the assistance of the New Jersey State Archives in Trenton, New Jersey; the Camden County Historical Society in Camden, New Jersey; the Maryland State Archives in Annapolis, Maryland; the Baltimore City Archives in Baltimore, Maryland; the City Archives and the Free Library of Philadelphia, Pennsylvania; Our Lady of Victory Center, Bishop Dunne Catholic School, and the Diocesan Archives in Dallas, Texas; and the National Archives offices in Leavenworth, Kansas; Philadelphia, Pennsylvania; and San Francisco, California.

When court documents were unavailable or lost, I relied on newspaper accounts, most notably the Camden *Courier-Post* and the *Philadelphia Inquirer,* which had the most comprehensive stories about Sally Horner's abduction, rescue, and death between 1948 and 1952.

I conducted dozens of interviews for the book, including several conversations with Sally's niece, Diana Chiemingo; one telephone conversation with Diana's father, Al Panaro, in 2014; two telephone conversations with Carol Taylor, in 2016 and 2017; and two conversations with "Madeline La Salle," in 2014. Other invaluable sources with firsthand memories of principal characters included "Rachel Janisch" and "Vanessa Janisch"; Fred Cohen and Peggy Braveman; Tom Pfeil; and Emma DiRenzo.

For the Nabokov sections, I relied on files, clippings, note cards, and letters deposited at the Library of Congress as well as at the Berg Collection, New York Public Library. Grateful acknowledgment for permission to access the Berg Collection is given to the Wylie Agency, on behalf of the Nabokov Estate, and to Isaac Gewirtz, Lyndsi Barnes, Joshua McKeon, and Mary Catherine Kinniburgh for their assistance and advice.

I also drew from the earlier work of Brian Boyd, Stacy Schiff, Andrew Field, Alexander Dolinin, and other Nabokov scholars. Schiff also generously shared her time, and advice, in a telephone conversation in April 2017, while Boyd was similarly helpful in an email exchange that same month. A telephone conversation with Walter Minton, in addition to earlier and later quotes, proved helpful with respect to the publication process of *Lolita* in the United States.

For additional historical context on Camden, I relied on *Camden After the Fall: Decline and Renewal in a Post-American City* by Howard Gillette (University of Pennsylvania Press, 2006) and the Local History: Camden website maintained by Phil Cohen at http://www.dvrbs.com.

ABBREVIATIONS

Berg: Vladimir Nabokov Archives, Berg Collection, New York Public Library, New York, NY

LOC: Vladimir Vladimirovich Nabokov Archives, Library of Congress, Washington, DC

NJSA: New Jersey State Archives, Trenton, NJ

VNAY: Brian Boyd, *Vladimir Nabokov: The American Years* (Princeton University Press, 1991)

Unless noted otherwise, all interviews were with the author.

INTRODUCTION: "HAD I DONE TO HER . . . ?"

6 no "little deadly demon": *Lolita*, p. 15.

6 It happened to the writer Mikita Brottman: Mikita Brottman, *The Maximum Security Book Club: Reading Literature in a Men's Prison*, pp. 196–197.

8 "I hate tampering with the precious lives": Nabokov, *Lectures on Russian Literature*, p. 138.

8 "It is strange, the morbid inclination": Nabokov, *Nikolai Gogol*, p. 40.

9 three increasingly tendentious biographies: The level of acrimony in Andrew Field's *VN* (1986) compared with *Nabokov: His Life*

in Art (1967) and *Nabokov: His Life in Part* (1977) is astonishing; the falling-out between biographer and subject would make an excellent play.

9 A two-part definitive study: Boyd's *Vladimir Nabokov: The Russian Years* (1990) and *VNAY* (1991).

10 Stacy Schiff's 1999 portrayal: Schiff, *Véra (Mrs. Vladimir Nabokov)*.

10 lifted its fifty-year restriction: Finding Aid, Vladimir Vladimirovich Nabokov Papers, Manuscript Division, LOC.

11 an earlier Nabokov story, "Spring in Fialta": *The Stories of Vladimir Nabokov*, p. 413.

ONE: THE FIVE-AND-DIME

15 Sally Horner walked into the Woolworth's: "Camden Girl Saved from Kidnapper in Calif," Camden *Courier-Post*, March 22, 1950, p. A1.

15 on a March afternoon in 1948: From Camden County prosecutor Mitchell Cohen's remarks at an April 2, 1950, court hearing, reported by the *Courier-Post* on April 3, p. A1.

16 A slender, hawk-faced man: Associated Press, March 22, 1950, taken from the *Lima* (Ohio) *News*, p. 5.

16 A scar sliced his cheek: Draft registration card, January 1944.

17 suicide of her alcoholic husband: Death certificate of Russell Horner, March 24, 1943.

17 Her homeroom teacher, Sarah Hanlin: *Philadelphia Inquirer*, March 23, 1950, p. 3.

17 Emma DiRenzo, one of Sally's classmates: Interview with Emma DiRenzo, November 13, 2017.

19 The telephone rang: Camden *Courier-Post*, March 23, 1950.

20 Ella let her concerns slide: United Press, *Salt Lake Tribune*, August 6, 1948, p. 5.

TWO: A TRIP TO THE BEACH

21 Robert and Jean Pfeffer were newlyweds: This section draws almost entirely from two newspaper reports that quoted Robert Pfeffer at length: Camden *Courier-Post*, March 24, 1950, p. 2; and *Philadelphia Inquirer*, March 24, 1950, p. 3.

24 Ella was relieved: Camden *Evening Courier*, August 6, 1948, p. 1.

24 Detective Joseph Schultz: *Courier-News*, Bridgewater, NJ, August 6, 1948, p. 15.

25 the lodging house: The 203 Pacific Avenue address came from the 1940 census; La Salle was known to return to addresses where he had lived in the past.

25 "He didn't take any of his or the girl's clothes": *Philadelphia Inquirer*, March 23, 1950, p. 1.

26 Marshall Thompson led the search: Camden *Courier-Post*, March 23, 1950, p. 1.

26 only six months before he'd abducted Sally: Mitchell Cohen's court statement, April 2, 1950.

THREE: FROM WELLESLEY TO CORNELL

27 The year 1948 was a pivotal one: This chapter largely draws upon *VNAY*, pp. 129–135, as well as letters reprinted in Nabokov, *Selected Letters: 1940–1977*.

28 Nabokov had also traveled: Itemized road trip summaries available at "Lolita, USA," compiled by Dieter E. Zimmer, http://www.d-e-zimmer.de/LolitaUSA/LoUSNab.htm.

28 "lovely, trustful, dreamy, enormous country": *Lolita*, p. 176.

28 "Beyond the tilled plain": Ibid., p. 152.

28 marriage to Véra was once again stable: *VNAY*, p. 129.

28 had been ill: Letter from Vladimir Nabokov to Katharine White, May 30, 1948.

29 "quiet summer in green surroundings": *VNAY*, p. 131.

29 "wrinkled-dwarf Cambridge flatlet": Ibid.

30 "ends with a feeling of hopelessness": Ibid.

30 Nabokov appreciated Wilson's gift: Letter from Vladimir Nabokov to Edmund Wilson, June 10, 1948, *Dear Bunny, Dear Volodya: The Nabokov-Wilson Letters, 1940–1971*, ed. Simon Karlinsky, p. 178.

30 "I was always interested in psychology": Field, *VN: The Life and Art of Vladimir Nabokov*, p. 212.

FOUR: SALLY, AT FIRST

34 Her legal name: Sally Horner's birth certificate, issued by the State of New Jersey Department of Health, April 18, 1937, obtained from the Department of Health office in May 2017.

34 When the subject came up: Interview with Diana Chiemingo, August 2014, and again in July 2017.

34 William Ralph Swain: Susan Panaro's birth certificate listing Swain as her father, State of New Jersey Department of Health, November 1926, obtained from the Department of Health office in May 2017.

35 One subject they all fretted about: Interview with Diana Chiemingo, July 2016.

35 That's where Ella met Russell Horner: *Asbury Park Press*, December 9, 1935, p. 9, and June 9, 1936, p. 7.

35 As for Russell Junior: Social Security application, February 1937.

36 Susan remembered the beatings: Interviews with Al Panaro and Diana Chiemingo, August 2014.

36 She took Susan and Sally to Camden: Camden telephone directory, 1946.

36 Russell became itinerant: Interview with Al Panaro, 2014.

36 He lost his driver's license: "'Short Cut' Costs Autoist License," *Philadelphia Inquirer*, March 27, 1942, p. 27.

36 By the beginning of 1943: *Asbury Park Press*, March 26, 1943, p. 2.

36 Later, when it became necessary: Camden *Courier-Post*, March 22, 1950, p. 1.

37 Her mother, Susannah: Obituary for Susannah Goff, *Trenton Times*, October 31, 1939; obituary for Job Goff, *Asbury Park Press*, January 12, 1943.

37 Susan, by now sixteen: Interview with Diana Chiemingo, July 2016; interview with Al Panaro, August 2014.

37 she and Al wed in Florence: Marriage certificate, NJSA.

FIVE: THE SEARCH FOR SALLY

39 Robert and Jean Pfeffer: *Philadelphia Inquirer*, March 24, 1950, p. 3.

40 At first Marshall Thompson worked: Joseph S. Wells, "Sleuth Closes Books on Tireless Search," Camden *Courier-Post*, March 22, 1950, p. 9.

40 He had been promoted to detective: "Marshall Thompson," DVRBS.com, http://www.dvrbs.com/people/CamdenPeople MarshallThompson.htm, accessed January 16, 2018.

41 "local pugilists": Camden *Courier-Post*, January 2, 1928.

41 His musical ability was called out: Camden *Courier-Post*, November 3, 1939.

41 "quantity of sugar and cream": Camden *Courier-Post*, March 22, 1950, p. 9.

SIX: SEEDS OF COMPULSION

46 "Of the nineteen fictions": Martin Amis, "Divine Levity," *Times Literary Supplement*, December 23, 2011.

46 suggested a more likely culprit: Roper, *Nabokov in America*, p. 150.

46 "an ape in the Jardin des Plantes": Nabokov, "On a Book Entitled Lolita," *Anchor Review*, 1957 (subsequently reprinted in the Putnam edition of *Lolita* and every edition since).

47 Nabokov supplementing his writing income: Beam, *The Feud*, p. 16.

47 The short story includes: "A Nursery Tale," reprinted in *The Stories of Vladimir Nabokov*, pp. 161–172.

48 features the so-called demonic effect: "Lilith," *Poems and Problems* (McGraw-Hill, 1969), reprinted in *Selected Poems* (Knopf, 2012), p. 84.

49 A paragraph in *Dar*: Nabokov, *The Gift*, pp. 176–177.

50 When Germany declared war: VNAY, p. 13.

50 "laid up with a severe attack": Nabokov, "On a Book Entitled Lolita."

51 "How can I come to terms": Nabokov, *The Enchanter*, p. 1.

53 "comparable to the one afforded": Simon Karlinsky, "Nabokov's Life and Lolita's Death," *Washington Post*, December 14, 1986.

54 As he later explained: Interview with Nabokov by Alfred Appel, *Wisconsin Studies in Contemporary Literature* 8 (1967).

54 Henry Lanz was a Stanford professor: VNAY, p. 33; Roper, *Nabokov in America*, p. 140.

55 Nabokov, however, denied it: Field, *Nabokov: His Life in Part*, p. 235.

SEVEN: FRANK, IN SHADOW

58 A likely birth date: La Salle's age was reported variably between fifty-two and fifty-six in 1950; his death certificate lists his birth date as May 27, 1896, and his Social Security application in 1944 lists May 27, 1895.

58 Frank Patterson and Nora LaPlante: Names listed on 1943 prison

intake form, NJSA. Different parental names, as well as hometowns, appeared on La Salle's Social Security application.

58 He said he served . . . prison has no record: Prison intake form, 1943; conversations with Greg Bognich, archivist, National Archives, Kansas City, KS.

58 every time he changed aliases: "Police Record of Girl's Abductor," Camden *Courier-Post*, March 22, 1950, p. 1.

58 It is as Fogg that a sharper picture: *News Journal* (Wilmington, Delaware), August 3, 1937, p. 24.

59 He met her at a carnival: *Philadelphia Inquirer*, August 3, 1937, p. 3.

59 Which they did: Cecil County marriage license of Dorothy May Dare and Frank La Salle, July 31, 1937, obtained from the Maryland State Archives.

59 Dorothy's father . . . was livid: *Philadelphia Inquirer*, August 4, 1937, p. 2.

60 "He told me the truth": *Philadelphia Inquirer*, August 3, 1937.

60 The next morning, La Salle appeared: *Philadelphia Inquirer*, August 4, 1937.

60 La Salle was fined fifty dollars: *Philadelphia Inquirer*, August 12, 1937, p. 2.

60 arrested La Salle on bigamy charges: Camden *Courier-Post*, March 22, 1950, p. 1.

61 Dorothy sued Frank for desertion: Ibid.

61 Three Camden police officers: Camden *Courier-Post*, March 25, 1950, p. 6.

62 The five girls: Names taken from Dorothy Dare's divorce petition against Frank La Salle, *La Salle v. La Salle*, Superior Court of New Jersey, 151-246-W127-796 (1944).

62 Sergeant Wilkie swore out a warrant: Camden *Courier-Post*, March 25, 1950, p. 6.

63 La Salle pleaded not guilty: Court docket, NJSA.

64 Dorothy and Madeline had moved: Interview with "Madeline La Salle," August 2014; *La Salle v. La Salle*, Superior Court of New Jersey.

64 La Salle was paroled: Camden *Courier-Post*, March 22, 1950, p. 1; prison intake form, 1950, NJSA; draft registration card, June 29, 1944; Social Security application, June 28, 1944.

65 a forged $110 check: Camden *Courier-Post*, March 22, 1950.

65 La Salle returned to Trenton State Prison: Ibid.; prison intake form, 1946, NJSA.

EIGHT: "A LONELY MOTHER WAITS"

67 She'd found work as a seamstress: *Philadelphia Inquirer*, December 10, 1948, p. 1.

69 The case had taken on added urgency: March 17, 1949, indictment date mentioned in subsequent reports by the *Philadelphia Inquirer* and Camden *Courier-Post*, March 22, 1950.

69 Dorothy Forstein's disappearance: "Kidnapping Story Spurs Search for Wife of Forstein," *Philadelphia Inquirer*, October 23, 1949, p. 1.

70 The Friday night after Dorothy vanished: "Reward Offered for Clue to Wife," Camden *Courier-Post*, November 17, 1949, p. 24.

71 Dorothy was declared legally dead: "Lost Wife Ruled Dead," *Philadelphia Inquirer*, October 15, 1957, p. 23.

71 Ella had difficulty sleeping: "A Tree Grows, a Lonely Mother Waits," *Philadelphia Inquirer*, December 10, 1948, p. 1.

NINE: THE PROSECUTOR

73 Mitchell Cohen was appointed: Obituary of Mitchell Cohen, *New York Times*, January 10, 1991.

73 did not have enough major crime: Gillette, *Camden After the Fall*, p. 25.

73 state party's de facto leader: Camden *Courier-Post*, November 7, 1950, p. 3.

73 many jobs he held in law enforcement: Obituary of Mitchell Cohen, *Philadelphia Inquirer*, January 10, 1991.

74 In his bespoke suits: Interview with Fredric Cohen, November 2017.

74 He'd met Herman Levin: Camden *Courier-Post*, June 1, 1956, p. 2.

74 Cohen also became a theatrical producer: Ibid.; also "Music Fair Opens to 1500," Camden *Courier-Post*, June 4, 1957, p. 1.

75 early in his tenure: "Make Up in Court," *Philadelphia Inquirer*, June 24, 1938, p. 19.

75 the murder of Wanda Dworecki: Account draws from coverage in the Camden *Courier-Post* and the *Philadelphia Inquirer*, as well as *State of New Jersey v. Dworecki*, January 11, 1940, and Daniel Allen Hearn, *Legal Executions in New Jersey: A Comprehensive Registry* (McFarland, 2005), pp. 376–377.

79 Shewchuk was paroled in 1959: "Pastor Put to Death in 1940," Camden *Courier-Post*, July 11, 2000, p. 6.

79 the death of twenty-three-year-old Margaret McDade: Account draws from coverage in the Camden *Courier-Post* and wire reports from the Associated Press, United Press, International News Service, and more.

80 Howard Auld did not die: "Auld Dies Tonight as Final Pleas of Mercy Are Denied," Camden *Courier-Post*, March 27, 1951, p. 1.

TEN: BALTIMORE

84 six-year-old June Robles: Obituary of June Robles, *New York Times*, October 31, 2017, supplied the bulk of details for this section.

86 Stan visited her parents: See Christine McGuire and Carla Norton, *Perfect Victim* (Arbor House/Morrow, 1988), for a complete account of the Colleen Stan case.

86 Dugard's eighteen-year bond: See Jaycee Dugard, *A Stolen Life* (2011); Elizabeth Smart, *My Story* (2013); and Amanda Berry and Gina DeJesus, *Hope* (2015), for further information on these cases.

87 had taken a taxicab: Mitchell Cohen statement, as reported by the Camden *Courier-Post*, April 3, 1950.

87 Sally later said: Camden *Courier-Post*, March 22, 1950, p. 1.

88 around West Franklin Street: Several addresses on this street were listed in an affidavit included with *State of New Jersey v. Frank La Salle*, A-7-54 (1954).

89 rape became a regular occurrence: Camden *Courier-Post*, March 22, 1950, p. 1.

89 To enroll Sally at Saint Ann's: Affidavit included with *State of New Jersey v. Frank La Salle*, A-7-54 (1954).

89 Sally got used to the new name: *Philadelphia Inquirer*, March 23, 1950, p. 1.

90 Breakfast in hand: "GEM F: On the Road to Hell," from Mary Reardon, *Catholic Schools Then and Now* (Badger Books, 2004), of-

fered a contemporaneous account of an elementary school student in the 1940s that proved helpful in imagining a typical day for Sally Horner during this time frame.

ELEVEN: WALKS OF DEATH

93 Camden believed in its own prosperity: Gillette, *Camden After the Fall*, p. 38.

94 At eight o'clock that morning: This account of the Howard Unruh massacre is drawn from several sources, including Seymour Shubin, "Camden's One-Man Massacre," *Tragedy-of-the-Month*, December 1949; Meyer Berger, "Veteran Kills 12 in Mad Rampage on Camden Street," *New York Times*, September 7, 1949; ". . . He Left a Trail of Death," Camden *Courier-Post*, September 7, 1974; Patrick Sauer, "The Story of the First Mass Murder in U.S. History," Smithsonian.com, October 14, 2015.

95 For Marshall Thompson: "Marshall Thompson," DVRBS.com.

97 Ferry had just finished: "John J. Ferry," DVRBS.com, http://www.dvrbs.com/people/CamdenPeople-JohnFerryJr.htm.

97 "When the other cops started arriving": Camden *Courier-Post*, September 7, 1974. Reproduced at http://www.dvrbs.com/people/CamdenPeople-HowardUnruh.htm.

99 Cohen walked over to the police station: Ibid.

100 He died in 2009 . . . the last survivor: Obituary of Howard Unruh, *New York Times*, October 19, 2009; obituary of Charles Cohen, Camden *Courier-Post*, September 9, 2009.

TWELVE: ACROSS AMERICA BY OLDSMOBILE

101 Nabokov finished the 1948–1949 academic year: This account of Vladimir Nabokov's whereabouts in the summer of 1949 is almost entirely drawn from *VNAY*, pp. 136–144.

106 "coach in French and fondle in Humbertish": *Lolita*, p. 35.

106 "white-frame horror": Ibid.

108 "You're a detestable, abominable, criminal fraud": Ibid., p. 96.

THIRTEEN: DALLAS

112 The journey from Baltimore to Dallas: Calculated via Google Maps.

112 However they traveled: Affidavit of Nelrose Pfeil, included with *State of New Jersey v. La Salle*, A-7-54 (1954).

112 The park was designed like a horseshoe: Interview with Tom Pfeil, November 2017.

112 La Salle had changed their names again: *Philadelphia Inquirer*, March 23, 1950.

112 The trailer park was owned: Interview with Tom Pfeil, November 2017.

113 He also enrolled Sally: Reproduction of a report card included with *State of New Jersey v. La Salle*, A-7-54 (1954).

113 Our Lady of Good Counsel no longer exists: See "Our Lady of Good Counsel, Oak Cliff," https://flashbackdallas.com/2017/10/01 /our-lady-of-good-counsel-oak-cliff-1901-1961/.

114 Her neighbors thought Sally: Affidavits from Nelrose Pfeil, Maude Smillie, Josephine Kagamaster, included with *State of New Jersey v. La Salle*, A-7-54 (1954).

115 She'd suffered an appendicitis attack: Camden *Courier-Post*, March 23, 1950.

FOURTEEN: THE NEIGHBOR

117 Ruth Janisch and her family: Interviews with "Vanessa Janisch," 2015 and 2016, and "Rachel Janisch," May 2017.

119 her second husband, Everett Findley: Marriage license of Ruth Douglass and Everett Findley, 1936.

119 She met husband number three: The 1940 census recorded Ruth, Findley, and Janisch all residing at the same home.

120 George and Ruth ran off: Marriage license, October 24, 1940.

121 "He never let Sally": Camden *Courier-Post*, March 27, 1950, p. 1.

121 El Cortez Motor Inn: *Philadelphia Inquirer*, March 22, 1950, p. 1; corroborated by a 1960 listing of motor courts obtained at the California Room, San Jose Public Library, July 2017.

122 Police in uniform shorts: "We'll Take the High Road," American Road Buildings Association, 1957, available at https://www.youtube .com/watch?v=wnrqUHF5bH8.

122 The friend told Sally: Camden *Courier-Post*, March 22, 1950, p. 1; *Philadelphia Evening Bulletin*, April 2, 1950.

FIFTEEN: SAN JOSE

125 On the morning of March 21: Camden *Courier-Post,* March 22, 1950, p. 1; *Philadelphia Inquirer,* March 22, 1950, p. 2; and other newspaper accounts.

126 Her brother-in-law, Al Panaro: Interview with Al Panaro, August 2014.

127 Hornbuckle had been elected: Howard Hornbuckle scrapbook, pp. 85–86, California Room, San Jose Public Library; Clerk-Recorder's Office, Santa Clara County Archives, Santa Clara, CA.

128 "Please get me away from here": *Philadelphia Inquirer,* March 22, 1950.

129 She started at the beginning: Camden *Courier-Post,* March 22, 1950.

130 La Salle elaborated his alternate reality: FBI report on Frank La Salle, memorandum prepared by agents C. Darrin Marron & Charles J. Prelsnik.

131 Ella Horner was overjoyed: Ibid.; also *Central New Jersey Home News,* March 22, 1950, p. 8.

131 Later that day: "Sally's Mother 'Relieved,' Admits She Was 'Foolish,'" Camden *Courier-Post,* March 22, 1950, p. 1.

132 "machina telephonica and its sudden god": *Lolita,* p. 205.

132 "all a-jitter lest delay": Ibid., p. 110.

133 "Give me some dimes and nickels": Ibid., p. 141.

133 "At the hotel we had separate rooms": Ibid., p. 142.

SIXTEEN: AFTER THE RESCUE

135 La Salle was charged: The White-Slave Traffic Act, or the Mann Act, is a U.S. federal law, passed June 25, 1910 (ch. 395, 36 Stat. 825; codified as amended at 18 U.S.C. §§ 2421–2424).

135 On the morning of March 22: Camden *Courier-Post,* March 23, 1950.

136 Commissioner Marshall Hall presided: "Sex Criminal Held as Girl Makes Charges Against Him," *San Bernardino County Sun,* March 23, 1950, p. 1.

136 When police officers attempted: "La Salle Held Under Mann Act," *Morning News* (Wilmington, Delaware), March 24, 1950, p. 1.

137 Even if he raised the full $10,000 bond: Camden *Courier-Post,* March 24, 1950, p. 1.

138 Back in Camden: Taken from Sally's statement reported by the Camden *Courier-Post,* March 22, 1950, p. 1.

138 A Camden grand jury: Camden *Courier-Post*, March 24, 1950, p. 1;
     Philadelphia Inquirer, March 24, 1950, p. 1.

139 Cohen, Dube, and Thompson flew: "Cohen Flies to Calif. to Re-
     turn La Salle on Kidnap Charge," Camden *Courier-Post*, March 27,
     1950, p. 1.

139 On Thursday, Sally was released: "Sally Meets Mother Again Af-
     ter 21 Mos.," Camden *Courier-Post*, April 1, 1950, p. 1.

139 Ella waited at the airport: "Sally's Mother Waited a Long Time to Hold
     Kidnaped Daughter Again," Camden *Courier-Post*, April 1, 1950, p. 2.

SEVENTEEN: A GUILTY PLEA

143 Frank La Salle wasn't allowed: Camden *Courier-Post*, March 30,
     1950, p. 1.

143 The solution was to transport La Salle: "Kidnap Victim Will Fly
     Home Tomorrow," *Oakland Tribune*, March 30, 1950, p. 51.

144 Mitchell Cohen was at the train station: Camden *Courier-Post*,
     March 31, 1950, p. 1.

144 La Salle, Detective Thompson, and Detective Dube: *Philadelphia
     Inquirer*, March 30, 1950, p. 2.

144 *City of San Francisco*: Extrapolated from sample train timetable,
     Union Pacific Railroad Company, Omaha, NE, Union Pacific Railroad
     Time Tables, April 1948, *The Cooper Collection of US Railroad History*.

144 New York–bound *General*: Times corroborated at American-Rails
     .com, https://www.american-rails.com/gnrl.html.

144 the trio of men: Camden *Courier-Post*, March 31, 1950, p. 1.

144 Mitchell Cohen told the press later on Sunday: Camden *Courier-
     Post*, April 2, 1950, p. 1.

145 Cohen arrived at the jail: "La Salle Given 30 Years," Camden
     Courier-Post, April 3, 1950, p. 1.

146 Judge Palese asked Cohen: Quoted text taken from *State v. Frank
     La Salle*, 19 N.J. Super. 510 (1952).

148 Because Frank La Salle pleaded guilty: Camden *Courier-Post*,
     April 4, 1950, p. 1.

EIGHTEEN: WHEN NABOKOV (REALLY) LEARNED ABOUT
SALLY

151 Vladimir Nabokov spent the morning: *VNAY*, pp. 146–147.

151 "I have followed your example": Letter from Nabokov to Katharine White, March 24, 1950, reprinted from *Selected Letters: 1940–1977*, p. 98.

152 But as Nabokov told James Laughlin: Letter from Nabokov to James Laughlin, April 27, 1950, ibid., p. 99.

152 described in their diary: Diary entry, November 17, 1958.

152 Robert Roper . . . was certainly convinced: Email to the author, August 25, 2016.

153 "will be given a choice": *Lolita*, p. 151.

153 "Only the other day we read": Ibid., p. 150.

154 Nabokov scholar Alexander Dolinin: "Whatever Happened to Sally Horner?," *Times Literary Supplement*, September 9, 2005.

155 "the stealthy thought": *Lolita*, p. 204.

NINETEEN: REBUILDING A LIFE

157 "When she went away she was a little girl": Camden *Courier-Post*, April 1, 1950, p. 2.

157 a family outing to the Philadelphia Zoo: Film clip provided by Diana Chiemingo, with permission.

158 "She has a definite ambition": *Philadelphia Inquirer*, March 29, 1950, p. 3.

159 Ella opted for a compromise: Interview with Al Panaro, August 2014.

162 "they looked at her as a total whore": Interview with Carol Taylor, August 2017.

162 "She had a little bit of a rough time": Interview with Emma DiRenzo, November 2017.

163 Sally found refuge in the outdoors: Interview with Al Panaro, August 2014.

TWENTY: *LOLITA* PROGRESSES

165 Vladimir and Véra left Ithaca: This chapter is largely drawn from *VNAY*, pp. 200–206; see also "Nabokov's Summer Trips to the West" at http://www.d-e-zimmer.de/LolitaUSA/LoUSNab.htm.

166 "Silly situation . . . to be smitten": letter from Véra Nabokov to Doussia Ergaz, September 2, 1951.

166 The Nabokovs changed their itinerary: *VNAY*, pp. 217–221.

TWENTY-ONE: WEEKEND IN WILDWOOD

169 Carol Taylor no longer remembers: Interviews with Carol Taylor, December 2016 and August 2017.

170 Edward John Baker drove down: *Vineland Daily Journal*, August 20, 1952, p. 1.

171 He died in 2014: Obituary of Edward Baker, *Vineland Daily Journal*, July 28, 2014.

172 "She impressed me as a darn nice girl": *Vineland Daily Journal*, August 20, 1952, p. 1.

173 Ed Baker pulled onto the highway: "Crash at Shore Kills Girl Kidnap Victim," Camden *Courier-Post*, August 18, 1952, p. 1; "Victim of 1948 Kidnaping Killed," *Morning News* (Wilmington, Delaware), August 19, 1952, p. 1.

173 The trip from Wildwood to Vineland: *Vineland Daily Journal*, August 20, 1952, p. 1.

173 Just after midnight on Monday: *Wildwood Leader*, August 21, 1952, p. 4; "W'bine Crews at 4-Vehicle Crash Scene," *Cape May County Gazette*, August 21, 1952, p. 1.

174 The death certificate: Unredacted copy of Sally Horner's death certificate obtained from NJSA.

174 The damage to her face: Interview with Al Panaro, August 2014.

174 Carol Starts was woken up: Interview with Carol Taylor, December 2016.

TWENTY-TWO: THE NOTE CARD

177 Vladimir Nabokov opened up a newspaper: Geographic location from *VNAY*, pp. 217–219.

177 The handwritten card reads as follows: Reproduced from LOC.

178 As Alexander Dolinin explained: Dolinin, "Whatever Happened to Sally Horner?"

180 "a golden-skinned, brown-haired nymphet": *Lolita*, p. 288.

181 Rather, he writes: Dolinin, "Whatever Happened to Sally Horner?"

181 how much damage he has caused: *Lolita*, p. 285.

182 Véra's diary note: Page-a-Day Diary, 1958, Berg.

182 "charming brat lifted from an ordinary existence": Letter to Nabokov from Stella Estes, quoted in *VNAY*, p. 236.

182 why Nabokov himself ranked Lolita: Page-a-Day Diary, September 17, 1958.

182 letter from Vladimir Nabokov to Shelley Estes.

TWENTY-THREE: "A DARN NICE GIRL"

183 a front-page interview with Edward Baker: "Vineland Youth, Bewildered by Publicity, Describes Sally Horner as 'Darn Nice Girl,'" *Vineland Daily Journal*, August 21, 1952, p. 1.

185 After he was treated: Camden *Courier-Post*, August 18, 1952, p. 1; "Driver Held at Shore in Horner Girl's Death," Camden *Courier-Post*, August 20, 1952, p. 11.

185 not Baker's first car accident: *Vineland Daily Journal*, July 24, 1951, p. 2.

185 Sally Horner's funeral: "Private Burial Held for Sally Horner," Camden *Courier-Post*, August 22, 1952, p. 4.

185 Emleys Hill Cemetery in Cream Ridge: See https://www.findagrave.com/memorial/11035529.

185 For Carol Starts, the funeral was awful: Interviews with Carol Taylor, December 2016 and August 2017.

186 Frank La Salle made his presence known: Interview with Al Panaro, August 2014.

186 The first court hearing: "Vineland Youth Freed in $1000 Bond Following Fatal Crash Near Shore," *Vineland Daily Journal*, August 19, 1952, p. 1; "One Fined in Fatal Crash," *Cape May County Gazette*, August 28, 1952, p. 4; September session, Cape May County Court (Criminal), September 3, 1952, pp. 19–21.

186 The most serious charge: *The State v. Edward J. Baker*, Indictment No. 283, New Jersey Superior Court, Cape May County, September 3, 1952.

187 The following week: September session, Cape May County Court (Criminal), September 10, 1952, pp. 25–26; "2 Plead Not Guilty in Girl's Death," Camden *Courier-Post*, September 12, 1952, p. 10; "Motorist Held in Death of Two," *Morning News* (Wilmington, Delaware), September 15, 1952, p. 12.

187 Carol was called to testify: Interview with Carol Taylor, August 2017.

187 Judge Tenenbaum threw out the charge: January session, Cape May County Court (Criminal), January 15, 1953, p. 63.

187 He faced a cluster of civil actions: "Civil Trials Set to Begin Before Jury," *Cape May County Gazette*, May 14, 1953, p. 1; "$115,800 Damage Suits Settled Out of Court," Camden *Courier-Post*, May 22, 1953, p. 15.

188 The byzantine nature of the lawsuits: "Fatal Accident Suits Resume After Mistrial," *Cape May County Gazette*, May 21, 1953, p. 1.

188 A new hearing lasted two days: Camden *Courier-Post*, May 22, 1953, p. 15; "Consolidated Trial Suits Settled," *Cape May County Gazette*, May 28, 1953, p. 2.

188 Written beside his name: Minutes, Cape May County Court (Criminal), June 30, 1954, p. 213.

TWENTY-FOUR: LA SALLE IN PRISON

189 a writ of habeas corpus: United States District Court for the State of New Jersey, C 679-50, "In the Matter of the Application of Frank La Salle for a Writ of Habeas Corpus," December 14, 1950.

190 Hughes was so incensed by La Salle's lies: "Kidnaper Seeking His Release from N.J. State Prison," Camden *Courier-Post*, September 21, 1951, p. 1.

190 He kept on, in a lengthy series: *State of New Jersey v. La Salle*, Superior Court of New Jersey, A-7-54 (1955).

192 Tom Pfeil denied she'd ever said: Interview with Tom Pfeil, June 2017.

193 his mother's supposed statement: *State of New Jersey v. La Salle*, Superior Court of New Jersey, A-7-54 (1955).

194 Frank La Salle also wrote letters: Interview with "Vanessa Janisch," May 2017.

195 Her mother, Dorothy: Obituary, Camden *Courier-Post*, August 2011.

196 Madeline did not learn any details: Interview with "Madeline," August 2014.

197 He appealed his sentence: *State of New Jersey v. Frank La Salle*, Superior Court of New Jersey, A-343-51 (1961).

197 He died of arteriosclerosis: Death certificate, State of New Jersey Department of Public Health.

TWENTY-FIVE: "GEE, ED, THAT WAS BAD LUCK"

199 another sensational crime: "Charge Is Due Today in 'Perfect Murder,'" *New York Times,* September 2, 1952, p. 17.

199 this case got an entire paragraph: *Lolita,* p. 287.

200 The G. Edward Grammer case: Case summary is derived from *State v. George Edward Grammer* (Transcripts), George E. Grammer, 1952, Box 1 No. 3544 [MSA T 496-67, 0/2/2/39], as well as subsequent appeals, including *Grammer v. State* (1953) and *Grammer v. Maryland* (1954). The entire case file is deposited at the Maryland State Archives, Annapolis, MD.

203 openly critical of mystery novels: Catherine Theimer Nepomnyaschy, "Revising Nabokov Revising the Detective Novel: Vladimir, Agatha, and the Terms of Engagement," The Proceedings of the International Nabokov Conference, March 24–27, 2010, Kyoto, Japan. Available at http://www.columbia.edu/cu/creative/epub/harriman /2015/fall/nabakov_and_the_detective_novel.pdf.

203 called out Dostoevsky as a hack: Nabokov, *Lectures on Russian Literature,* p. 109—while this line is the opinion of the author, Nabokov's judgment "Let us always remember that basically Dostoeveski [sic] is a writer of mystery stories" is meant to be pejorative.

203 As Véra told their close friend Morris Bishop: Schiff, *Véra (Mrs. Vladimir Nabokov),* p. 232.

203 stabbing murders of Dr. Melvin Nimer and his wife: Nabokov almost certainly read "Prosecutor Says Boy, 8, Confesses Killing Parents; Boy Said to Admit Killing Parents," *New York Times,* September 11, 1958, p. 1.

204 police detectives still claiming as recently as 2007: "Nimer Now" (video), *Staten Island Advance,* February 11, 2007, http://blog.silive .com/advancevideo/2007/02/nimer_now_458.html.

204 venture west one more time: *VNAY,* pp. 223–226.

TWENTY-SIX: WRITING AND PUBLISHING *LOLITA*

205 Vladimir Nabokov wrote a note: Page-a-Day Diary, 1953, Berg.

205 "a novel I would be able to finish": Letter from Nabokov to Edmund Wilson, June 15, 1951.

206 "crumpling each old manuscript sheet": *VNAY,* p. 225.

206 "enormous, mysterious, heartbreaking novel": Letter from Nabo-
    kov to Katharine White, September 29, 1953.

206 when Nabokov wrote to Edmund Wilson: Letter from Nabokov to
    Edmund Wilson, 1947.

206 The first time was in the fall of 1948: Schiff, *Véra (Mrs. Vladimir
    Nabokov)*, p. 166.

207 "Véra came to the rescue": Roper, *Nabokov in America*, p. 149.

207 "one day in 1950": Interview with Nabokov by Herbert Gold, *Paris
    Review* 41 (Fall 1957), reprinted in Nabokov, *Strong Opinions*, p. 105.

207 *Lolita* was ready to be submitted: *VNAY*, pp. 255–267.

208 Edmund Wilson read half: Letter from Edmund Wilson to Nabo-
    kov, November 30, 1954.

208 grew "negative and perplexed": Letter from Mary McCarthy to
    Nabokov, November 30, 1954.

208 Wilson's present wife, Elena: Letter from Elena Wilson to Nabo-
    kov, November 30, 1954.

209 parody piece in the *New Yorker*: Dorothy Parker, "Lolita," *New
    Yorker*, August 27, 1955, p. 32.

209 Nabokov joked to Edmund Wilson: Letter from Nabokov to Ed-
    mund Wilson, February 19, 1955.

209 founder and publisher of Olympia Press: Account is drawn in
    large part from John De St. Jorre, *Venus Bound: The Erotic Voyage of
    the Olympia Press and Its Writers*.

209 submitted *Lolita* to Girodias: *VNAY*, p. 265.

210 As Nabokov later recalled: "*Lolita* and Mr. Girodias," *Evergreen
    Review* 45 (1967), reprinted in Nabokov, *Strong Opinions*.

211 Nabokov received a letter from Walter Minton: Letter to Nabokov
    from Walter Minton, August 30, 1957, reprinted in *Selected Letters:
    1940–1977*, pp. 224–225.

211 had succeeded his father, Melville: "Walter Minton on the House
    'Lolita' Built," *New Yorker*, January 8, 2018.

212 "I thought Nabokov had": "The Lolita Case," *Time*, November 17, 1958.

212 he had all but given up: Letter from Nabokov to Walter Minton,
    December 23, 1957.

212 *Lolita* had been banned in France: *VNAY*, pp. 310–315.

212 Minton's letter augured a change: Letter from Nabokov to Walter

Minton, September 7, 1957; letter from Véra Nabokov to Minton, September 19, 1957; De St. Jorre, *Venus Bound*, p. 144.

213 "'Don't ever open your mouth'": Undated interview with Walter Minton by John De St. Jorre, quoted in *Venus Bound*. When I spoke to Minton in August 2017, he brought up the legality of *Lolita*'s copyright status without prompting: "I still wonder about that damn copyright."

213 As Minton explained: Inference from Nabokov letters to Walter Minton, January–February 1958.

214 Vladimir and Véra Nabokov left Ithaca: *VNAY*, pp. 362–364.

214 "Vladimir was a tremendous success": Page-a-Day Diary, August 1958, Berg.

214 Minton sent the following telegram: Reprinted in *Selected Letters: 1940–1977*, p. 257.

215 Elizabeth Janeway's rave review: "The Tragedy of Man Driven by Desire," *New York Times Book Review*, August 17, 1958.

215 The reorder number from retailers: *VNAY*, p. 365.

215 "ought to have happened thirty years ago": Letter from Nabokov to Elena Sikorski, September 6, 1958.

216 The indefinite leave of 1958: *VNAY*, p. 378.

TWENTY-SEVEN: CONNECTING SALLY HORNER TO *LOLITA*

217 Peter Welding was a young freelance reporter: Obituary of Peter Welding, *New York Times*, November 23, 1995.

217 Welding remembered reading of Sally's plight: "Lolita Has a Secret, Shhh!," *Nugget*, vol. 8, no. 5, November 1963.

222 a *New York Post* reporter named Alan Levin: Obituary of Alan Levin, *New York Times*, February 17, 2006.

224 The Nabokovs subscribed: Manuscript box, miscellaneous clippings, 1960–1965, Berg.

226 Schiff . . . strongly advised against reading: Interview with Stacy Schiff, April 2017.

TWENTY-EIGHT: "HE TOLD ME NOT TO TELL"

229 Decades after Ruth Janisch: Account is largely drawn from interviews with "Rachel Janisch," May 2017, and "Vanessa Janisch," March 2015, March 2016, and May 2017.

TWENTY-NINE: AFTERMATHS

235 Ella had connected with a new partner: 1951 Camden telephone directory records both residing at 944 Linden Street.

236 made their union legal: California marriage records, 1965, retrieved through Ancestry.com.

236 Five years later, Burkett was dead: Death certificate, State of California Department of Public Health, 1970.

236 Diana didn't learn the truth: Interview with Diana Chiemingo, August 2014.

237 Ella settled back in New Egypt: Obituary of Ella Horner, 1998, Ancestry.com.

237 Susan died in 2012, and Al passed away: Obituary of Susan Panaro, *Burlington County Times*, August 5, 2012; obituary of Al Panaro, KoschekandPorterFuneralHome.com, February 25, 2016.

237 "Did you say that Sally Horner": Interview with Carol Taylor, December 2016; email from Robin Lee Hambleton, November 2017.

238 Edward Baker got on with his life: Obituary of Edward Baker, *Vineland Daily Journal*, July 28, 2014.

238 On the afternoon of Wednesday, May 17: *Vineland Daily Journal*, May 18, 2007.

238 The two Camden police detectives: "Wilfred L. Dube," DVRBS.com, http://www.dvrbs.com/people/CamdenPeople-WilfredLDube.htm; "Marshall Thompson," DVRBS.com.

239 Howard Hornbuckle served one more term: Obituary of Howard Hornbuckle, *Petaluma* (California) *Argus-Courier*, May 9, 1962, p. 4.

239 Mitchell Cohen's health suffered: Camden *Courier-Post*, August 30, 1950, p. 1.

239 including . . . the 1955 execution: Camden *Courier-Post*, May 4, 1955, p 1.

239 Then came his next career move: Obituary of Mitchell Cohen, Camden *Courier-Post*, January 1991.

239 Palese eventually acquiesced: Obituary of Rocco Palese, Camden *Courier-Post*, February 27, 1987, p. 19.

239 Cohen served three years: Obituary of Mitchell Cohen, *Asbury Park Press*, January 9, 1991, p. 8.

240 Véra Nabokov continued: Page-a-Day Diary, 1958, Berg.

241 went out to dinner at Cafe Chambord: The account largely draws from Véra Nabokov's November 26, 1958, entry in ibid.

241 unduly preoccupied with a *Time* magazine article: "The Lolita Case," *Time,* November 17, 1958.

241 unbylined but written by . . . Joyce Haber: Haber worked at *Time* as a researcher and reporter from 1958 through 1966. While Minton did not comment on whether he had a relationship with Haber, a former colleague recognized the writing as Haber's.

244 Comedians turned *Lolita* into late-night fodder: *VNAY,* p. 375.

244 Another bizarre stunt: Ibid., pp. 415–416.

245 "by nature I am no dramatist": Preface to *Lolita: A Screenplay,* p. ix.

245 changed his mind about adapting *Lolita*: Letter from Nabokov to Morris Bishop, *Selected Letters: 1940–1977,* p. 309.

247 "a graceful ingenue but not my idea": Nabokov, "On a Book Entitled Lolita," *Novels, 1955–1962,* p. 672.

247 "I didn't have to play Lolita": Interview in the *New York Times,* 1971.

247 European newspapers: Manuscript box, miscellaneous clippings, 1960, Berg.

248 "a first-rate film with magnificent actors": *VNAY,* p. 466.

248 gave his approval for the musical: Ibid., p. 583.

249 "I think he's crude": Interview with Nabokov by Robert Hughes, WNET, September 2, 1965, reprinted in Nabokov, *Strong Opinions.*

EPILOGUE: ON TWO GIRLS NAMED LOLITA AND SALLY

255 The irritation is evident: Interview with Nabokov, BBC, July 1962, reprinted in Nabokov, *Strong Opinions,* p. 15.

255 He denied Humbert Humbert: *Paris Review,* "The Art of Fiction No. 40," 1967.

256 After one stern denial: BBC interview, reprinted in *Strong Opinions,* p. 17.

256 "with crooning sounds and fancy endearments": Nabokov, *Speak, Memory,* p. 49.

257 "The desperate truth of *Lolita*'s story": Nafisi, *Reading Lolita in Tehran,* p. 33.

BIBLIOGRAPHY

SELECTED WORKS BY VLADIMIR NABOKOV

Laughter in the Dark (1932; first published in English as *Camera Obscura*, 1936; second, revised English-language edition published in 1938; third edition published in 1963).

Despair (1934; translated into English in 1937; second English-language edition published in 1965).

The Gift (1938–1952, translated into English and published in 1963).

The Enchanter (written in 1939, published posthumously, translated and with a preface by Dmitri Nabokov, 1986).

Nikolai Gogol (1944).

Speak, Memory (1951, as *Conclusive Evidence;* revised edition published in 1966).

Lolita (1955).

Pnin (1957).

Pale Fire (1962).

The Annotated Lolita (edited with preface, introduction, and notes by Alfred Appel, Jr., 1970; revised and updated, 1991).

Strong Opinions (1973).

Lolita: A Screenplay (1973).

Lectures on Literature (edited by Fredson Bowers, introduction by John Updike, 1980).

Lectures on Russian Literature (edited and with an introduction by Fredson Bowers, 1981).

Vladimir Nabokov: Selected Letters, 1940–1977 (edited by Dmitri Nabokov and Matthew J. Bruccoli, 1989).

Dear Bunny, Dear Volodya: The Nabokov-Wilson Letters, 1940–1971 (edited by Simon Karlinsky, 1979; reprinted 2001).

Letters to Véra (edited and translated by Olga Voronina and Brian Boyd, 2015).

WORKS BY OTHERS ON NABOKOV

Alfred Appel, Jr., and Charles Newman, *Nabokov: Criticism, Reminiscences, Translations, and Tributes.* Northwestern University Press, 1970.

Alex Beam, *The Feud: Vladimir Nabokov, Edmund Wilson, and the End of a Beautiful Friendship.* Pantheon, 2016.

Brian Boyd, *Vladimir Nabokov: The Russian Years.* Princeton University Press, 1990.

———, *Vladimir Nabokov: The American Years.* Princeton University Press, 1991.

———, *Stalking Nabokov: Selected Essays.* Oxford University Press, 2012.

Mikita Brottman, *The Maximum Security Book Club.* Harper, 2016.

Andrew Field, *Nabokov: His Life in Art.* Little, Brown, 1967.

———, *Nabokov: His Life in Part.* Viking, 1977.

———, *VN: The Life and Art of Vladimir Nabokov.* Crown, 1986.

John De St. Jorre, *Venus Bound: The Erotic Voyage of the Olympia Press.* Random House, 1994.

Michael Juliar, *Vladimir Nabokov: A Descriptive Bibliography.* Garland Publishing, 1986.

Michael Maar, *The Two Lolitas.* Verso, 2005.

———, *Speak, Nabokov.* Verso, 2010.

Azar Nafisi, *Reading Lolita in Tehran.* Random House, 2003.

Ellen Pifer, ed., *Vladimir Nabokov's* Lolita: *A Casebook.* Oxford University Press, 2003.

Andrea Pitzer, *The Secret History of Vladimir Nabokov.* Pegasus, 2013.

Robert Roper, *Nabokov in America.* Bloomsbury USA, 2015.

Phyllis Roth, ed., *Critical Essays on Vladimir Nabokov.* G. K. Hall, 1984.

Stacy Schiff, *Véra (Mrs. Vladimir Nabokov).* Random House, 1999.

Marianne Sinclair, *Hollywood Lolita: The Nymphet Syndrome in the Movies.* Plexus, 1988.

Russell Trainer, *The Lolita Complex.* Citadel, 1965.

Graham Vickers, *Chasing Lolita: How Popular Culture Corrupted Nabokov's Little Girl All Over Again.* Chicago Review Press, 2008.

Michael Wood, *The Magician's Doubts: Nabokov and the Risks of Fiction.* Princeton University Press, 1997.

Lila Azam Zanganeh, *The Enchanter: Nabokov and Happiness.* W. W. Norton, 2011.

ARTICLES AND WEBSITES

Anonymous (attributed to Joyce Haber), "The Lolita Case." *Time,* vol. 72, no. 20, November 17, 1958.

Martin Amis, "Lo Hum and Little Lo." *The Independent,* October 24, 1992.

Brian Boyd, "The Year of Lolita." *New York Times Book Review,* September 8, 1991.

Robertson Davies, "Mania for Green Fruit." *Saturday Night,* October 11, 1958.

Alexander Dolinin, "Whatever Happened to Sally Horner?: A Real Life Source of Nabokov's *Lolita.*" *Times Literary Supplement,* pp. 11–12, September 9, 2005.

Leland de la Durantaye, "The Pattern of Cruelty and the Cruelty of Pattern in Vladimir Nabokov." *Cambridge Quarterly,* October 2006.

——, "Lolita in *Lolita,* or the Garden, the Gate and the Critics." *Nabokov Studies* 10 (2006).

Sarah Herbold, "(I Have Camouflaged Everything, My Love): *Lolita* and the Woman Reader." *Nabokov Studies* 5 (1998–1999): 81–94.

Elizabeth Janeway, "The Tragedy of a Man Driven by Desire." *New York Times Book Review,* August 17, 1958.

Landon Jones, "On the Trail of Nabokov in the American West." *New York Times,* May 24, 2016, https://www.nytimes.com/2016/05/29/travel/vladimir-nabokov-lolita.html.

Erica Jong, "*Lolita* Turns Thirty: A New Introduction." *New York Times Book Review,* June 5, 1988.

Vladimir Nabokov, "On a Book Entitled Lolita." *Anchor Review,* 1957. (Reprinted in every English-language edition of *Lolita* since 1958.)

Heine Scholtens, "Seeing Lolita in Print." Thesis for M.A. Programme in Book History, Leiden University, 2005 (uploaded in 2014).

Delia Ungureanu, "From Dulita to Lolita." In *From Paris to Tlön: Surrealism as World Literature.* Bloomsbury Academic, 2017.

Dieter Zimmer, "Lolita, USA." 2007. http://www.d-e-zimmer.de/Lolita USA/LoUSpre.htm.

Dieter Zimmer and Jeff Edmunds, "Vladimir Nabokov: A Bibliography of
    Criticism." 2005. http://www.libraries.psu.edu/nabokov/forians.htm.

OTHER SOURCES

Amanda Berry and Gina de Jesus, *Hope: A Memoir of Survival in Cleveland.*
    Viking, 2015.

Phil Cohen, "Local History—Camden, NJ." http://www.dvrbs.com.

Jeffery M. Dorwart, *Camden County, New Jersey: The Making of a Metropoli-
    tan Community, 1626–2000.* Rutgers University Press, 2001.

Jaycee Dugard, *A Stolen Life.* Simon & Schuster, 2011.

Howard Gillette, Jr., *Camden After the Fall.* University of Pennsylvania
    Press, 2005.

Michelle Knight, *Finding Me: A Decade of Darkness, A Life Reclaimed.*
    Weinstein Books, 2014.

Elizabeth Smart, *My Story.* St. Martin's Press, 2013.

PERMISSIONS

IMAGES

A candid shot of Sally holding a newspaper: Panaro/Chiemingo family
    archives

Véra and Nabokov chasing butterflies: Carl Mydans, The LIFE Picture
    Collection/Getty Images

Carol Starts, Sally Horner's best friend, summer of 1952: Panaro/Chi-
    emingo family archives

Edward Baker's high school graduation photo, 1950: Vineland High
    School yearbook, retrieved via Ancestry.com

AP story of Sally Horner's death transcribed onto a note card by Vladi-
    mir Nabokov: LOC note card, Box 2, Folder 14, Vladimir Vladimirov-
    ich Nabokov Papers, Manuscript Division, Library of Congress,
    Washington, DC

Image accompanying Peter Welding's November 1963 article for *Nugget*:
    courtesy of the author

Sally Horner, age fifteen, summer of 1952: Panaro/Chiemingo family
    archives

Frank La Salle, prison intake photo: National Archives, Kansas City office

INDEX

Note: Page numbers in *italics* indicate photos.